Electric Universe

ALSO BY DAVID BODANIS

The Body Book

The Secret House

Web of Words

The Secret Garden

The Secret Family

$E=mc^2$

Electric Universe

How Electricity Switched On the Modern World

DAVID BODANIS

THREE RIVERS PRESS • NEW YORK

Copyright © 2005 by David Bodanis

All rights reserved.
Published in the United States by Three Rivers Press,
an imprint of the Crown Publishing Group,
a division of Random House, Inc., New York.
www.crownpublishing.com

Three Rivers Press and the Tugboat design are registered trademarks
of Random House, Inc.

Originally published in hardcover in the United States by Crown Publishers,
an imprint of the Crown Publishing Group,
a division of Random House Inc., New York, in 2005

Library of Congress Cataloging-in-Publication Data
Bodanis, David.
Electric universe : how electricity switched on the modern world /
David Bodanis.
p. cm.
Originally published: New York : Crown Publishers, c2005.
Includes bibliographical references and index.
1. Electricity.   2. Force and energy.   I. Title.
QC522.B64   2006
537—dc22   2005052916

ISBN-13: 978-0-307-33598-2
ISBN-10: 0-307-33598-4

Printed in the United States of America

Design by Helene Berinsky

10 9 8 7 6 5 4 3 2 1

First Paperback Edition

To Sam and Sophie, my beloved children

To the city of Chicago, where what wisdom I've been able
to grant them long ago began

And to Natasha, who taught me—when I needed it most—
how to navigate without a map

CONTENTS

"Mysterious affair, electricity."

—SAMUEL BECKETT, "Theatre II"

INTRODUCTION

When my father was a little boy, in a village in Poland before the First World War, an electricity blackout wouldn't have been especially important. There were no cars, which meant there were no traffic lights to fail, and there were no refrigerators— just blocks of ice or cool rooms—so food wouldn't suddenly spoil either. A very few rich people would find their electric lights failing if the generators stopped working in their homes, and the single telegraph line that passed through the town might stop operating, but by and large daily life would continue as it had before.

By the time my father's family had migrated to Canada, and then to Chicago in the early 1920s, a big power outage would have been different. People would still have been able to buy things—there were no credit cards that depended on computer verification—but the streetcars that workers rode to the factories wouldn't run. The telephones that offices depended on wouldn't work either, and the skyscrapers that the city was so proud of would quickly have become inaccessible, or at least their upper floors would have, as their elevators failed too. It still wouldn't have been a complete catastrophe. Farm crops could still be raised—there weren't many tractors—and coal-fired trains and steam-driven ships would have kept the city pretty well supplied.

Today, though? I live in London now, where people can be pretty phlegmatic, but I still wouldn't want to be around for a complete blackout. Most radios and TVs plug in these days, so it would be difficult to find out whether your kids' school was still open. Your cell phone might still operate, but with no way of recharging your battery you'd be pretty careful about using it. Driving the kids to school on the off chance it was open would be too much of a gamble, for gas stations depend on under-ground storage tanks, and until the blackout ended, stations wouldn't be able to use their electrically operated pumps to bring up more fuel to sell to anyone in the city. You couldn't stock up on groceries—no credit cards working—nor could you get more cash, for ATMs depend on electrically run computers too.

Within a week the city would really have broken down. Police stations would be isolated with their phones not working, and pretty soon their radio batteries would lose their charge as well; no one could call ambulances, for their radios or phone links would be out too. A few people might try walking to hospitals, but there wouldn't be much there: no X rays, no refrigerated vaccines, no refrigerated blood, no ventilation, no lighting.

Going to the airport to try to escape wouldn't help, for with backup generators not working, the airport's radars would have shut down, nor could planes take off on manual control, for any fuel that remained in underground tanks couldn't be pumped up. As the blackout spread, the nation's ports would have closed, with no electricity to run the cranes that moved their large containers and no way to check electronic inventories. The military might try to guard fuel convoys, but with their own vehicles running low on fuel, that wouldn't last long. If the blackout was worldwide, isolation would intensify. The Internet and all e-mail would have gone down very quickly; next the phone lines; finally, the last television and radio broadcasts would end.

Starvation would probably begin in the dense cities of Asia, especially with no air conditioning at food warehouses; within a few weeks of a complete blackout almost all the world's cities and suburbs would be unlivable. There would be fighting, pretty desperate, for food and fuel, and with a world population of 6 billion, few people would have a chance of surviving.

But what if it were not only our supply of electricity that failed? What if the very existence of electrical forces stopped? All the Earth's oceans would gush upward and evaporate as the electrical bond between water molecules broke apart. DNA strands within our body would no longer hold together. Any air-breathing organism that was still intact would begin to suffocate, for without electrical attraction, the oxygen molecules in air would bounce uselessly off the hemoglobin molecules in blood.

The ground itself would open up and begin to melt as the electrical forces that hold the silicates and other substances of our planet together let go. Mountains would collapse into the voids left where the continental plates had torn apart. In the last moments, a few living beings would see the sun itself switch off, as our star's electrically carried light abruptly stopped and the world's very last day turned to night.

Why doesn't any of this happen? The force of electricity is very powerful, and has been operating nonstop for more than 13 billion years. But it's also utterly hidden, crammed deep within all rocks and stars and atoms. The force is like two Olympian arm-wrestlers, whose struggle is unnoticed because their straining hands barely move. There are almost always equal amounts of positive and negative electric charge within everything around us—so well balanced that although their effects are everywhere, their existence remains unseen.

For long eons it remained like this—as galaxies evolved and planets formed, as continents and trees and grasses appeared on Earth. Occasionally there must have been brief sightings in the past. Our australopithecine ancestors would have noticed abrupt bursts of lightning, as would early humans. But as soon as it appeared, this force would quickly have returned to the invisible realm from which it had come. For most of history, humans simply stumbled around it, unaware.

In one of his writings, Isaac Bashevis Singer imagines a peasant in medieval Ireland who takes off his flaxen tunic one night and notices bright sparks leaping from the fabric. If Singer's peasant called in the village priest and other wise elders to see it happen the next night, they would be unlikely to notice anything: static electric sparks usually appear only in dry air, and Ireland is wet. No one would believe what he'd seen; no one would have been able to examine it further. Even in dry desert countries, dust or sand could make scattered sparks seem to appear and disappear in purely random ways.

There were many fragmentary efforts to penetrate this hidden world, from classical Greek times on, but even into the mid-1700s there was little true knowledge. The breakthrough came with the work of a conveniently vain Italian investigator, Alessandro Volta, in the 1790s. It would be a great honor, he felt, to locate the portal from which this mysterious "electricity" emerged, and after much effort he realized where he should search. He found that if he pressed a coin-shaped copper disc against one side of his tongue, and a zinc disc against the other, and then touched the tips of the two coins together, a tingling sensation would race across his tongue. He'd located the world's first steadily operating "battery"—in his mouth.

Volta soon found that any two metals would do the trick so long as they were separated by a small amount of saliva, brine, or other corrosive liquid. He didn't know why this worked, or how the brine was making extra electrons appear on one of the metals, but he could send the tingling spray of electrons along a laboratory bench through a wire and was getting famous just for describing it, and that was enough. The stuff that came out of this battery rushed forward like water in a river, so it was called an "electric current."

As the Victorian era dawned, that was still most of our knowledge: two metals, when positioned near each other, could

sometimes produce a sparking current within a wire connecting them. It seemed a weak, merely curious phenomenon. But it
was the first useful door into a world that had been sealed and
hidden.

In this book I show what has happened in the two centuries
since humankind opened that door, which took a mere two centuries. The first part looks at the Victorian researchers who had
only a few tenuous glimpses of electricity, yet created devices
never before imagined. There were telephones and telegraphs
and lightbulbs; roller coasters and fast streetcars—and ever more
electric motors powering them all. There was even an electrical
fax machine operating efficiently in France in 1859—*before* the
American Civil War.

The world started to change. The new wave of electrical technologies helped lead to the modern corporation and to votes for
women, to suburbs stretching far from cities, and tabloid newspapers, and, influenced by crisp telegraph messages, to a new
Hemingway–style prose. One exuberant telephone executive
apparently remarked that Americans had become the first people who would interrupt sex to take a phone call.

Things might have stopped there, but in the mid-1800s two
of England's greatest scientists opened the door to electricity's
domain a crack wider. They found that the electricity sizzling
through wires doesn't move along on its own. There's something else, a pervasive rush of unseen waves, that pushes it
along. In the second part of the book, I show that all the space
around us—from the air above to our flesh within—is filled with
millions of such flying, invisible waves.

Several researchers who heard of these waves were so awed by this knowledge that they became more confirmed than ever in their religious faith; others saw in these hitherto unsuspected waves a mechanism by which extrasensory perception and other psychic phenomena could take place.

The idea that our world is permeated by invisible waves was so strange that it took a giant engineering project, deep under the cold waters of the Atlantic, to begin to convince the majority of researchers that it was true. Before the nineteenth century was out, a determined experimentalist found ways to release these waves from inside copper wires and send them flying free. This discovery led to the first experiments with cell phones (and a primitive mobile phone was operating on London's Portland Place in 1879, outside the present-day BBC Broadcasting House). A few decades afterward, television and radar took shape as well, hitching a ride with the same invisible waves. In the second and third parts of the book, I show how those waves were used: first for peace, then for war.

In the twentieth century the door opened even further. A few physicists were finally able to look directly at the face of electricity. The younger ones were awed by what they saw; many of the older ones—including even the great Einstein—pulled back, saying that what was now revealed was something they could never accept.

What the researchers had found was that the atoms inside us don't really look like miniature solar systems, with electrons orbiting like miniature planets around a tiny sun. Rather, these electrons—which are central to how electricity affects us—can wildly teleport from one location to another. It was the only partially predictable nature of these jumps that Einstein was thinking of when he famously said, "God does not play dice with the universe." (And it was to that dictum that his friend Niels Bohr exasperatedly replied, "Einstein, stop telling God what to do!")

That jumping of electrons within us would be as if the Earth were an electron that could instantly shoot away from the sun and take up a position hovering above the planet Jupiter. Families having breakfast in Duluth, Minnesota, would look out their window, see an Extremely Large Red Spot, and groan as they grabbed hold of the table, knowing what was likely to happen next. As they held the plates down, they'd suffer through another series of jolts, as the Earth leaped elsewhere in space—possibly returning them to their original orbit, but possibly flinging them to a distant solar system for more adventures and shattered crockery.

These new findings might have remained just a laboratory curiosity, but the twentieth century brought the world's gravest wars, and in those times of emergency anyone who promised to explain how charged electrons moved stood a chance of receiving extra funds for his investigation. In the fourth part of the book I show how these wild teleporting jumps came to be used in the century's first huge thinking machine, and ultimately in the microchips now operating daily inside mobile phones, passenger jets, oil well pumps, and all the other devices that our imagined blackout showed are so important.

Electricity also operates in our own private thinking machines—the human brain—and the book ends with a section showing how this was discovered, and how it led to pills such as Prozac, which, when swallowed, actually turn into liquid electricity that can shape our very mood. We use electricity to sense every person and sight around us; anyone we touch or kiss is forever beyond our reach, with glowing electrons from our fingers or lips always getting in the way of direct contact.

It's a wondrous topic, but a complex one. To keep the main text from being loaded with details and qualifications, notes at the end of the book give more backing for the various explanations, as well as derivations of such curios as the fact that invisible waves stream out from our brains with a wavelength of about 200 miles; beyond that, there's a website—davidbodanis.com—that looks into the science and history a little further.

The stories along the way are as much about religion, love, and cheating as they are about impersonal science or technology. They take us from Hamburg cellars during a World War II firestorm to the mind of Alan Turing, brilliant computer inventor, hounded by the authorities of the very country he'd saved; from the slum-born Michael Faraday, slurred by his contemporaries because of his religious faith (yet who used his faith to become the first to see electric forces weaving invisibly through space); to a pampered artist, Samuel Morse, who eagerly ran for mayor of New York on a platform of persecuting Catholics, and who learned more about how telegraphs operate than he ever cared to admit, from a frontiersman who couldn't believe anyone would wish to patent such an obvious idea.

There's an exuberant twenty-something immigrant to America,

Alexander Bell, desperate to capture the love of a deaf teenage student, and there's the forty-something Robert Watson Watt, desperate to escape from a boring marriage and the tedium of 1930s Slough. There's Otto Loewi, who wakes up one Easter eve realizing he has solved the problem of how electricity works in our body, yet in the morning, agonizingly, can't read the scrawled explanations he jotted beside his bed during the night; there's the boy from rural Scotland, James Clerk Maxwell, who was treated as a fool for years by bullies at his elementary school, yet who became the nineteenth century's greatest scientific theorist, able to envision the inner structure of the universe in a way that scientists of a later era would realize was profoundly true. All of these stories illuminate how the immense force of electricity was gradually seen: how it was led out from its hidden domain—and what we, imperfect humans, have made of the enhanced powers it has granted.

PART I

WIRES

When the universe was very young, in the first moments after the Big Bang, powerful charged electrons began to pour out of the swirling furnace that filled empty space. Many became part of simple hydrogen atoms that tumbled through the cosmos and ended up within huge stars.

In their long sojourn within the stars, and then even more when the stars blasted apart, multitudes of those simple atoms were squeezed together with such force that larger atoms were created.

Metals such as copper, iron, and silver were born.

For eons these metals, too, floated through space. In time they fell toward a new solar system, and became part of ore deposits on the North American continent. They were joined by metal atoms that had been created in other distant starbursts. Hidden deep inside each atom, as the ore lay buried, powerful electron charges remained.

Mountains rose and fell. Giant reptiles hunted in fern forests; ecosystems changed, and giant mammals hunted in coniferous and broad-leaf forests. Small groups of arrow-using humans arrived from Asia; thousands of years later, more humans arrived, on giant floating vessels from Europe and Africa. There were cruel frontier wars, and new settlements arose. The soil was turned over for planting, and probed for metal ore. The hidden electrical charges, unchanged for billions of years, were about to be released.

1

The Frontiersman and the Dandy

ALBANY, 1830, AND WASHINGTON, D.C., 1836

Joseph Henry was a strapping, rawboned American from the distant reaches of the frontier state of New York, who by the age of thirty had left jobs as a handyman (too boring), a builder (too low paid), a metal worker (too hot), and then, most disastrously, a surveyor, where he'd let himself be talked into leading a crew for several winter months through the forests toward the Canadian border (far, far too cold). In 1826, freshly back from the surveying and miraculously undamaged by frostbite, he heard about a position at a school in his native town of Albany. The salary would be low, and as a new hire Henry would be stuck with teaching elementary arithmetic along with other topics. But the classrooms would be warm—so he nabbed the new job in an instant.

He had dozens of farm boys to keep quiet—boys expert at spitballs and pencil jabs—but he knew what would keep them happy. Boys like building stuff, and the bigger the better. He'd just give them something really, really big to build.

The new field of electricity would be a good place to hunt for an idea, he decided, for it was an area he'd long been intrigued by in a casual, self-educated way. The word covered a whole range of effects, from small sparks of static to the giant lightning bolts investigated, famously, by Philadelphia's best-known retired printer, Benjamin Franklin. It was interesting stuff, for it always seemed to involve a strange sort of sparking substance. No one knew much more about what it actually was, though—but Henry had just heard an intriguing hint.

Not many European science journals made it to Albany, but one that did, months late—after the usual long ocean voyage and the wait for the ice on the Hudson to melt—described the extraordinary experiment of a recently demobilized British artillery officer, William Sturgeon.

When Sturgeon had taken a piece of iron and wrapped a coil of wire around it, nothing much had happened. That was fair enough. But when he'd gone on and connected the wire to a battery so that Volta's mysterious "electric" current sparked through the wire, the ordinary iron had seemed to come alive. It had turned into a strong magnet that pulled other pieces of iron toward itself, as if an invisible force were jumping from the wire into the iron. Switch off the battery and everything stopped; the original lump of iron became inert, and what it had lifted toward itself fell off. It was no longer magnetized.

Sturgeon didn't know what to make of this discovery, but Henry knew exactly what to do with it. His boys were suffering a general winter restlessness, and this amazing toy is what he could use to keep their attention. The boys were good with their hands, and liked solid physical things, as their parents did too: most of the wooden houses around Albany were newly built, and it was often the settlers themselves who had done the building. If he could make the electrical magnet really huge, he'd have them all on his side.

Creating one of Volta's batteries to power the whole contraption wasn't too hard, for there was plenty of metal ore available, either locally or from the big ports to the East. Henry was a fast worker, and by 1827 he had duplicated Sturgeon's work, creating an electromagnet that could lift nine pounds. The boys in his classroom wrapped still more coils of wire around the lump of iron. The battery was switched on. The wrapped chunk of iron could now lift more than twenty pounds. Henry did yet more wrapping, and when the coils of wire got so close together that they touched and start crackling, he simply asked his new bride for her petticoats, enlisted her to help cut them into strips, and used the cloth to insulate the copper wires so they could be wrapped even more tightly together.

By 1830, there on the edge of the American frontier, he had succeeded in building a small electromagnet that could lift 750 pounds. The schoolkids loved it, and no doubt asked some of their friends—still out on the farms—to come by to watch this marvel. Henry was proud of it too, but for a deeper reason. He was a deeply religious man and had always suspected that God had created marvels not visible to ordinary eyes. With enough ingenuity, though, we could magnify and reveal God's hidden work.

Henry kept on going. Before he was done, he had wrapped a

small chunk of iron so tightly with wire that when the battery was switched on, that small chunk lifted more than 1,500 pounds—the weight of several big blacksmith anvils. Just so everyone could see, he hoisted the whole thing up on a sturdy scaffolding. Disconnect the battery, and with an almighty crash— "This never fails to produce a great sensation," Henry wrote—all that weight would come plummeting down.

There are times when it pays to worry about underlying explanations, and there are other times when good honest tinkering is the best way forward. Future researchers would learn that the atoms we're made of aren't solid little ball bearings, but that parts of them are electrically charged and can be torn off. Those torn-off bits are called electrons, and researchers by the end of the 1800s came to believe that they were what rolled forward inside a wire, and it was the power of those charged electrons that gave an electric current its strength. When an electric cable gets cut open in a storm, the sparks that spray out are a sign of the streams of electrons that were inside the cable. Within a phone wire, electrons are rolling forward, and inside a powerful searchlight, even more electrons are moving.

Henry himself would become a big part of the research leading up to those findings: in years to come he became recognized as one of the greatest American physicists of the nineteenth century, ending up as director of the Smithsonian. But at this point, fairly young and still stuck in Albany, he knew he wasn't going to arrive at much of an explanation of all the lifting and banging his electromagnets could produce.

Instead he invented the telegraph. It was easy enough. He just lengthened the wire that stretched from his battery to the electromagnet. The electrons that poured out from the metals in the battery were powerful enough to seemingly hurtle themselves along. Rather than keep the battery right next to the electromagnet when he turned it on, he could carry it to the next room, or down the hall, or even downstairs. The mysterious electric force would squirt along from his battery through the wire, and turn on the electromagnet waiting at the end of the wire. Any lump of iron next to it would be tugged close.

It would have been a pretty galumphy telegraph if he'd had it lift up and then drop giant chunks of metal every time he wanted to send a single letter of the alphabet. Instead, Henry went back to using a tiny electromagnet, even slimmer than the ones the British officer had created. Right next to it he put something resembling a small, clickable castanet, like a little metal tongue. Turn on the battery and a current ran down the wire, the electromagnet powered up, and it pulled the castanet toward itself. You heard a click. Turn off the battery and the electromagnet let go. You heard another click as the castanet went back. Henry realized it would be easy to communicate just by agreeing that different arrangements of clicks would represent different letters. The electric charges that had been dormant so long in the ancient metals was now coming out, to power this "click."

The Albany schoolboys loved the new invention, especially when Henry let them get rid of the castanet and use the ringer of a bell instead. When one of the kids switched the battery on and off, his friends in the next room or even down the hall would hear the ringer sound in short bursts, just as fast as his hand could move.

At this point a very different sort of individual entered the field, someone who would have his own ideas about how these electric discoveries could be used. Samuel Morse had studied fine arts at Phillips and Yale, and in his early twenties he was living off his parents' money in London. He seemed to be just another ethereal art student, explaining, in one of his many letters home, that he was asking for more money not because of his difficulties in painting a portrait that actually looked much like any of his sitters, but rather because—and one has to be honest here—he was too good for even the refined British to recognize: ". . . had I no higher thoughts than being a first-rate portrait-painter," he explained to his mother, "I would have chosen a far different profession. My ambition is to be among those who shall revive the splendor of the fifteenth century; to rival the genius of a Raphael, a Michelangelo, or a Titian. . . ."

But underneath that genteel pose, Morse was raving. His father, an evangelical Calvinist, had brought him up to believe that America was being destroyed by secret conspiracies, and when Morse returned to America and found he still couldn't make a living, he took his father's ideas and went further. The awful powers that were attacking America—and keeping deserving artists from commercial success—were ones he alone had now identified. There were Negroes and Jews and other undesirables, of course, but behind them all were the Catholics, and behind them were the Jesuits—secret, armed Jesuits, fanning out from their missions across the United States, storing guns in Irish nunneries, and all under the control of the Emperor of Austria.

When his pamphlets were ignored, he took it as a further sign of the conspiracy, so in 1836 Morse ran for mayor of New York, with straightforward slogans to persecute Catholics. "We have to resist the MOMENTOUS evil that threatens us," he wrote. "Will you not awake to the reality of your danger? Up! Up! I beseech you. To your posts!"

He lost, of course, and then he started sulking, and then—holed up in an isolated aerie in one of the highest buildings overlooking the recently founded New York University—he worked out what needed to be done. The Jesuits were controlling America through invisible forces, so it was necessary that Morse—and all other good Americans—develop a similar means of fighting back. Something that could stretch everywhere and flash information along at the speed of electricity would be just about ideal.

Luckily, on a ship voyage from London not long before, he had overheard one passenger discuss some of the ways electricity was being used for such long-distance contacts. It was pretty well known. Joseph Henry was teaching at the College of New Jersey (soon to be renamed Princeton University) by then, and some news of his work had been publicized. There also had been similar trials in Europe. Charles Wheatstone and William Cooke in England, for example, had stretched a wire from the train terminal at Euston in London to a train depot at a strange round building in Camden, over a mile away. When they connected the Euston end of the wire to a battery, an electromagnet in Camden got switched on. (The local residents loved it, for the single telegraph wire replaced the piercing whistle and whappingly loud drums that had been used to communicate arrivals and departures before.)

Now in his New York retreat, after struggling to make a telegraph of his own work efficiently—he suffered from as great a lack of ability in his mechanical tinkering as in his artistic efforts—Morse almost gave up in frustration. He was certain that there had to be an easier way, so he decided to get help from someone who actually knew how these mysterious electrical substances operated and could explain it to him.

Which is how, probably on a spring day in 1838, Joseph Henry found a surprisingly impassioned ex-painter at the door of his Princeton office.

Henry was as easygoing at Princeton as he had been in Albany. The students liked him. By now he was stretching telegraph cables for more than a mile around the Princeton campus, and students regularly helped him in the work. Henry had often declared that patents were the sort of thing that had held Europe back. He happily explained to Morse how his system worked—the batteries and the electromagnet and the spools of wire. In America, a young and growing country, it was right and proper, Henry believed, for all good citizens to share what they learned.

When Morse left Princeton, he knew what was best for one good citizen at least. He'd always been keen to patent anything he could—while still an artist he'd struggled with an impractical marble-carving device—and he patented the information he'd picked up from Henry's work, as well as techniques he'd learned from reading European reports.

The idea of using a simple code for the telegraph was already surprisingly widespread. The great German mathematician Carl Friedrich Gauss, who'd strung a telegraph along his Göttingen campus in 1833, had set up his receiving electromagnet to tug a needle either to the left or the right. If it shifted to the right, that meant a letter such as *e* was intended. If it shifted to the left, that could mean the letter *a*. Two shifts to the right might signify an *i*, while other combinations of left and right signified other letters, all the way to the least common letters, such as *x* and *z*. Other researchers regularly came up with similar codes, for it made sense to have the simplest signals stand for the most common letters, and more-complex signals stand for rarer letters. (It was possible to find which letters were most common simply by going to a print shop and looking. Typesetters tended to accumulate big boxes of type bearing the letter *e*, since they had to use them so often; they had only smaller boxes for the rarely demanded *q*, *x*, and *z*.)

It took Morse several years—and judicious financial involvement with key members of Congress—before he secured enough government funds actually to build a large, working prototype of his telegraph device. In its first week of commercial operation in 1844, connecting Washington with Baltimore, it took in just thirteen and a half cents in paid traffic, but in the next year an expanded line was taking in over a hundred dollars each week, and within a decade Morse was one of the wealthier men in North America.

Did it matter that he had largely stolen the idea for his invention? Telegraphs were already operating in England and Germany, and in America other inventors were close behind them. Someone else would no doubt have helped jump-start the American system if Morse hadn't done it.

Although divine justice didn't keep Morse from earthly riches, it did strike in another way. Joseph Henry had a satisfying life, at ease with his students and respected by his peers. Morse, however, having engaged in so much subterfuge, spent much of the next three decades stuck in litigation trying to defend the patents he'd railroaded through in his name. (There was one embarrassing moment when his lawyer was forced to announce to the Supreme Court that a notebook which honestly, truly, had Morse's original handwritten notes on the telegraph, had mysteriously disappeared in a fire to which there were no witnesses— shortly before it was supposed to be shown to the court.)

The invention had other unexpected consequences as well. Before the telegraph, a horseman carrying a message between two cities had to transport the bulk of perhaps one thousand pounds of animal over rocks, muddy ruts, and the occasional fallen tree. That took plenty of food for the two mammals involved, as well as the bulky technologies of saddle, stables, horseshoes, and the like. With few exceptions, information at the start of the nineteenth century traveled at about the same rate it had managed in ancient Sumer.

A thin copper telegraph wire, however, carrying the same message, merely had to move an electric current composed of those mysterious "sparks." No one in Henry's day knew that those sparks had something to do with electrons, which weigh less than a millionth of an ounce each. But his contemporaries did realize that what was in the wires was much smaller—and moved a lot faster—than anything in the ordinary world. A battery small enough to fit in a thimble could power a message great distances and at great speed.

The world changed. Financial news could now be sent instantly between cities, and—along with enhanced opportunities for insider trading—a new style of corporation arose. Offices in far distant cities could be easily linked. Train networks became more complex, for telegraphs strung beside the rail lines could synchronize departures and arrivals across entire countries.

There were psychological shifts as well. Before electricity became widespread, time had been something local, changeable, personal. New York and Baltimore, for example, had kept their clock systems several minutes apart, since they were at slightly different longitudes, and noon arrived a few minutes later in Baltimore than it did in New York. Each city was a separate world, so it was fair to think that each individual, strolling here or there, or working on this or that isolated farm, was part of a similarly separate world. But now those worlds could be synchronized and, wherever you were, you knew where you fit in the tight, universal "control" of clocked time.

It was an early form of globalization. As the telegraph spread into central and eastern Europe, millions of peasants were forced to take on last names, the better for newly enlarged government bureaucracies to educate, tax, or enlist them. In the past, the quick movement of mass armies had been possible only through the occasional genius of a Napoleon or the temporary enthusiasm of revolutionary citizens. By the mid–1800s, however, tens of thousands of bewildered military recruits regularly found that their marches or train transport, synchronized by telegraph, delivered them as close as possible to where thousands of armed enemy soldiers were, unfortunately, similarly being assembled.

Newspapers stopped being journals of slow discussion or courtly gossip, and started depending on featured foreign correspondents. Diplomatic crises had less time to calm down, as foreign–office lassitude was regularly broken by "urgent," just-arrived messages. Mass political movements sprang up faster than before; new factory techniques spread more quickly as well.

One more consequence: Ever more jobs were available in America, and with this telegraph–aided globalization, increasing numbers of Europeans were able to make the arrangements to move there. New steamship lines appeared to bring the workers over, first by the thousands, then by the tens of thousands. There were Jews and Protestants, but there were also Catholics, lots of them, bursting with energy. The result was a dynamic, immigrant–rich America. It was everything Joseph Henry loved.

It was everything Samuel Morse hated.

2

Aleck and Mabel

BOSTON, 1875

For several decades the steady transformation of the world continued as the consequences of the telegraph spread around the globe. But then, starting in the 1860s, a long pause began. Partly it was because one of the main centers of innovation, the United States, was consumed by an awful Civil War and its repercussions. But even into the 1870s, there was still no fundamentally new technology.

Wall Street and the City of London had plenty of money to fund new ideas, but only if they were easily understandable ideas, ones that merely modified what could already be done. Yet electricity's greater effects—the next power tapped from the vast number of charged particles hidden inside us—would come through creations that were hitherto unimagined. Only in the hot summer of 1875 did the first of such fundamentally new inventions appear, and it was the work of the most innocent of young men, a twenty-eight-year-old teacher who'd set up his own tutoring business in Boston. He was driven neither by avarice nor by the urge for power.

His invention came because he was in love.

Unfortunately, the object of his affections, Mabel Hubbard ("You do not know," he desperately wrote, "you cannot guess—how much I love you") was one of his students, and this meant he felt obliged to declare his interests to her parents first. But although he emphasized his fine prospects, and even made sure her parents had seen his flourishing signature, where an added *k* made his name Aleck and not the humdrum Alec, they were unimpressed. For Mabel was from a very rich family—her father owned much of downtown Boston—she was also barely seventeen, and, most important, she'd had scarlet fever when she was a child. The infection had spread to her ears, destroying her hearing. Aleck was a teacher of the deaf, and Mabel had been able to hear no sound or music or voice for more than ten years. She'd learned some signing and lip-reading, but lived a protected life. Aleck was firmly forbidden to declare himself.

Their first attempt to keep their courtship secret lasted about twenty-four hours, for Mabel's older sister decided to help things along by inviting this intriguing older teacher to the family house. She even left Aleck alone with Mabel in the garden for at least ten minutes before interrupting them to present a handful of flowers to play "he loves me, he loves me not."

After that episode, Mabel's parents gave Aleck another talking-to, and a few weeks later her mother read him a letter that, she explained, would finally prove that it was over. In it

Mabel seemed to be saying that she was not in love with her teacher, and that was that.

Aleck vowed he would respect the wishes of Mabel's parents, and his vow this time actually lasted all the way till August, when he tracked the family down at their summer home in Nantucket. On his first day at the Ocean House hotel on the island, there was a tremendous thunderstorm, and he stayed inside, pouring out his heart in a letter: "I have loved you with a passionate attachment that you cannot understand . . . it is for *you* to say whether you will see me or not." The next day he went to her house to deliver it, was met at the door by Mabel's cousin, and was told again that Mabel was not going to see him, that she didn't love him, that it was over.

It's said that blindness separates you from things, but deafness separates you from people. Aleck had been determined to bridge this separation—not just through words, for he could already communicate with Mabel, but in the contact of real love. He was disconsolate all the way back to Boston. But he was also bursting with ideas, and he had spent almost a year now coming close to a very good one indeed. For his full name was Alexander Graham Bell, and he was about to create the telephone.

By the early 1870s, when Aleck had arrived in America from Britain, it had seemed possible to send more than one signal over a standard telegraph line. (Think of someone drumming a fast pattern with his right hand, and a separate, slower pattern with his left hand. Each hand is sending a sequence of taps, and they overlap in time, but if you listen carefully you can distinguish the two patterns.) Aleck had been one of the inventors working on that, in fits and starts, but then he'd become distracted by the far stronger idea that he might be able to shunt entire sounds rather than mere clicks down a telegraph line. He built a prototype to try out that idea.

In March 1875—a half-year before the declaration to Mabel's family—he'd taken his apparatus to someone he'd heard had once been important in science, the now aged Joseph Henry, retired Princeton professor. It was an electric wire stretching out from a battery and connected to a single tuning fork. By switching the battery on and off, he could make the tuning fork hum in various ways. Aleck asked if he should develop this himself, or just let others go ahead. As he remembered years later: "I felt that I had not the electrical knowledge necessary to overcome the difficulties. [Henry's] laconic answer was—'GET IT.' I cannot tell you how much these two words have encouraged me."

Now, back in Boston, Aleck realized that he did have a chance to win Mabel's hand after all. What if he went ahead and finished his invention? There would be money and fame, respect from her parents, and—could they be so lucky?—a grand wedding bursting with flowers.

A handful of other inventors had reached this point, but none that Bell was aware of had been able to go further. Aleck's love, however, extended not just to Mabel but to the whole community of the deaf. It was a powerful motivation. His own mother had been unable to hear; that was what had led him to start his tutoring business in the first place. He had grown up in a family where understanding how a sound could be communicated was central to every daily task.

His grandfather had been an actor and elocution expert; George Bernard Shaw modeled Professor Henry Higgins in *Pygmalion* partly on his example. Indeed, Aleck's father had spent so much time helping his wife communicate that he'd realized that the usual way of classifying speech—simply making long lists of all possible sounds—was never going to help a deaf person.

Instead, Aleck's father had focused not on the final sounds but on the process of creating them. He drew a series of little diagrams showing different positions of the tongue and lips, which he called his Visible Speech kit. The diagrams were simple enough for even a child to read. To demonstrate this, Aleck's father had guests come over and challenged them to make sounds that were exotic or unexpected—a click from the Southern African Xhosa language; a rolled Spanish r; even a sneeze. Then he'd pull out the appropriate cards showing how to produce the sounds and pass them to his sons, who'd been kept out of the room. Relying only on the cards, Aleck and his brother would move their tongues and widen or narrow their throats to produce exactly the targeted sound.

This creation of sounds had always been Aleck's real interest, and at the school in Boston he now perfected all the ideas he needed to make a telephone that worked. One of his favorite pupils was Georgie Sanders, only five years old when Aleck began tutoring him. Aleck spent time just playing with him, and then began pasting little word labels on all Georgie's toys, showing him the word each time they played with that toy. Then, after a while: "I remember one morning [Georgie] came down stairs in high spirits, very anxious to play with his doll. . . .

I produced a toy-horse; but that was not what he wanted. A table; still he was disappointed. He seemed quite perplexed to know what to do, and evidently considered me very stupid. At last, in desperation, he went to the card-rack and after a moment's consideration, pulled out the word 'doll' and presented it to me." With that, the teaching took off: Aleck was making the boy connect the idea of playing with a doll, to the strange squiggle of letters D O L L on a piece of cardboard. Every afternoon after that, when it was time for Aleck to appear at the house, five-year-old Georgie would be at the window, eagerly awaiting his older friend.

What Aleck understood, through the time with Georgie, was something that most other researchers into possible telephones were missing. Typically those researchers linked a telegraph to dozens of tuning forks at the receiving end, sending a combination of clicks to try to make the right tuning forks vibrate and produce a word.

It never worked well. Aleck knew, from his patient hours with Georgie Sanders, that in communication you start with a thought, such as a five-year-old *really* wanting to play with a doll. Then from that thought you select the appropriate word— the carefully inked D O L L in Georgie's card rack—and only then do you break the word into sounds. Aleck knew that any telephone would have to follow this sequence. But who knew how hidden thoughts could be turned into audible vibrations?

Mabel did. She was in love with Aleck beyond measure. Her mother hadn't been telling the truth when she'd pretended to read from the "rejection" letter earlier in the summer.

"I think I am old enough now," Mabel wrote to her parents that year, "to have a right to know if [Aleck] spoke about [his feelings] to you or Papa. I know I am not much of a woman yet, but . . . it comes to me more and more that I am a woman such as I did not know before." And then, emphatically: "You need not write about my accepting or declining [his] offer."

She and Aleck had fallen in love, with neither admitting it, over a year before. They'd developed an intimacy during the months of teaching: she had found someone who would see through her deafness; he had found a woman who was confident enough to match all his interests. When she was late he'd meet her carriage and they'd run together through the snow to his schoolroom; they had talked about politics and families, and sometimes just gossiped. And time after time, as she tried different sounds, he had touched her throat, and she had touched his, quite properly, with other students around, ostensibly just for the sake of identifying how different words produced different vibrations. But each had suspected what the other was thinking.

"Your voice has a beautiful quality," he had whispered to her after one session, using hand gestures and careful enunciation so she could read his lips. She was startled and wrote about it to her family. She had almost no memory of what her voice was like, and she knew she'd never hear it again. No one had ever thought to tell her it was beautiful.

When her cousin had sent their visitor away that morning in Nantucket, Mabel was furious, and when she later got a letter through to Aleck, she felt her chances had been ruined. "Perhaps it is best we should not meet awhile now," she wrote, "and that when we do meet we should not speak of love." The resolution collapsed almost instantly, of course, and defeat for the parents was inevitable. Mabel's mother had already experienced life with one strong-willed daughter—Mabel's sister—and knew it was a losing proposition to try to put obstacles in the way of young love. She invited Aleck to the house to talk about his new ideas—they *did* seem so promising—and then she invited him again. There was at least one tempestuous meeting with Mabel's father, and the necessary intervals when the lovers sulked and then made up, until finally, on Thanksgiving Day 1875—Mabel's eighteenth birthday—she told Aleck that she loved him, and she kissed him, and she even agreed that she would marry him—so long as he was willing to make just one little change. It was simply to drop the final k from his name, and for the rest of his seventy years, that was what he did.

The understanding of how vibrations produced sounds had dominated his unspoken courtship with Mabel. It was a motif throughout his teaching. When some of the younger children at the Boston school were endangered by fast-racing horse-drawn wagons—which they couldn't hear coming—Alec had them try holding balloons in their bare hands when they walked outside. Vibrations from the unheard wagons would travel along Boston's cobblestones and make the balloons shudder, alerting the children to hurry aside.

In 1875, his love and his invention came together. Why not create a device with an *artificial* voice-box that mimicked a human one? He knew it could be done, for as a teenager he and his brother had created an artificial throat and lips: the tongue was made of several small coated paddles, and behind it they'd put the larynx from a dissected sheep, and below it they'd attached a bellows to serve as the lungs powering it all. By pumping the bellows and carefully moving the larynx and the homemade tongue and lips, they'd made it yelp "Mama!" so clearly that an upstairs neighbor called down that someone should feed the baby.

A little later they practiced on the family dog, a long-suffering Skye terrier. First they got the dog to produce a steady growling hum, and then—with copious biscuits as inducement—Aleck gently manipulated the Skye's larynx while his brother worked its lips. To their friends' amazement, several distinct simple "words" came out.

Now in Boston Mabel's father had also become resigned to his daughter's match, and decided to help Alec so he could support his daughter. He began paying for an assistant for Alec, a young machinist named Tom Watson, and together the two twenty-something hopefuls began by preparing a hand-sized sheet of stiff parchment. Hold it in front of your mouth, and as you poured out words, the parchment rippled back and forth in time with the sound vibrations from your throat—like the balloons that Alec's young charges had held, like the skin on Mabel's throat in those tremulous days of teaching months before.

To create a working telephone, Alec needed a way to turn the patterns that words made in the shaking parchment into electricity. He'd learned a lot in the tedious months he'd spent working to improve the telegraph, and he remembered one key observation. If the electric charges pouring out from a battery are poured into a wire, a current rolls along at a steady rate. But if you bend or twist the wire, the electric current can't pour through so easily. You've increased the resistance inside the wire.

Alec brought the parchment close to his mouth and placed a wire along the other side of it, almost touching the paper. Each time he spoke, the puff of air from his mouth pushed the parchment against the wire. It didn't bend it very much—the air vibrations when we speak are tiny—but for the even smaller streams of electric current pouring along inside the wire, it was enough. He imagined the sides of the wire starting to bend in; the electrical sparks or fluid that he pictured flowing inside it— he too didn't have a clear idea of what was inside a wire—was being squeezed so that less got through. The resistance was high. Then, when the giant Gulliver outside stopped speaking, the awful windblasts and earthquake ended. The wire straightened out; the full stream of electric current sped through it. The resistance was back to being low.

There would be many modifications in later years, but that's basically how a telephone works. You speak into a microphone that's a bit like a human larynx or voice-box. The microphone quivers from your voice's uneven air blasts—shaking rapidly with exuberant high sounds, but barely at all with silences or quiet *umms*. The quivering microphone leads a wire to send an electric current surging along in an exact copy of that uneven pattern.

When the current arrives in the listener's receiver, everything happens in reverse. All the up-and-down patterns your voice has produced in the microphone are now in the receiver. When a lot of current arrives, it makes a surface—in this case plastic—vibrate quickly, and the listener hears a loud, clear voice. If the electric current is weak, the membrane inside the earpiece vibrates slowly and the listener hears only a quiet whisper. With Bell's invention, even a whisper could be carried, undimmed, through thousands of feet of wire.

With some encouragement from Mabel's anxious parents, Alec applied for a patent, and then an improvement on the patent, and soon after there was the wedding, which, to Mabel's delight, had lilies, lots of them. Alec gave her pearls, a silver pendant in the shape of a telephone, and 1,497 shares of stock in the fledgling Bell Telephone company—which would be worth, if kept in the family, several billion dollars today. Less than a year later their first child was born; their marriage lasted till the end of their lives.

3

Thomas and J.J.

NEW YORK, 1878

B ell's work in the 1870s was the start of a great outpouring of new discoveries. A proconsul from the Roman Empire, suddenly transported to the muddy swampland of the American settlement of Fort Dearborn, in the year 1850 A.D.—a little before Bell's work—would not have been especially surprised at what he found. There were horse-drawn vehicles and wooden houses, and candles or oil lamps to hold back the night. The few telegraphs that might be found in big cities had scarcely changed the quality of daily life. But if that proconsul had returned a single lifetime later, in 1910, that muddy town would have exploded to become the city of Chicago—and amid the cars and electric lights and telephone poles, where powerful electric charges were led whirling along at immense velocities, our time-voyaging proconsul would have been utterly startled.

This second generation of transformations was begun by individual inventors such as Bell. But as the 1870s went on, an increasing number of discoveries were made by larger groups of researchers, working in a new style of industrial research laboratory. They were the ones who produced the generators and streetcars and motors and lighting systems that created modern Chicago and other great metropolises around the world.

Running these big research labs required a different personality than that of the gentle Alec Bell. The new research directors had to understand electricity, of course, but they also had to be willing to work on assignment . . . and not worry too much what those assignments were.

Thomas Edison was the most powerful of these new industrial research chiefs, and one of his great successes came in 1877 when he accepted an important assignment to crush Bell. The world's largest telegraph company, Western Union, had been watching what Bell was doing, and even before his final model was ready, they'd tried to get him to leave a prototype overnight at their New York headquarters so they could "examine" it. Bell was a trusting man, but not *that* trusting; he kept the prototype secure in his own hotel room.

Once he had his patent, more-direct measures were needed, for who was going to let an upstart undercut a giant industry? Certainly not William Orton, the head of Western Union. His strategy was almost embarrassingly simple. America after the Civil War was a violent place. Strikes were often resolved with rifles and dynamite; patents were stolen; fledgling investment houses were destroyed by established firms. It wasn't surprising that within the technology field, predators began to appear, generally bankrolled by rich financiers. When they identified a new electrical product, they would try to find a technological mercenary skilled enough to produce the same device using a slightly different process. The original inventor would be destroyed; the company that had arranged for the copy—and the mercenary who produced it—would become rich.

Because Bell's telephone threatened to undermine the entire telegraph business, Orton had to go to the most skilled enforcer he knew. This was the young Thomas Edison, a man who, as Orton happily explained to a friend, "had a vacuum where his conscience ought to be."

Edison was almost exactly Bell's age, but from a very different background. Instead of the doting parents and uncles and education in Scotland and London that Bell had, Edison had a father who had once whipped him in a public square, and he had left school in frontier Michigan when he was barely a teenager. He'd supported himself as an itinerant telegraph operator for years, sleeping in cheap hotels and rooming houses across America. This would have been hard enough for any fifteen-year-old, but Edison was also very hard of hearing. When he wanted to hear a piano properly, he'd have to get a piece of wood, bite down on it, and then push the wood as hard as he could against the piano. ("I haven't heard a bird sing since I was twelve years old," he once casually remarked.)

When he got married, young, he ended up with a woman with whom he soon found he had almost nothing in common; when he tried his first legitimate invention, a quick vote-counting machine for legislatures, he found that he was laughed at: everyone in the know understood that legislators did not want their votes to be counted quickly.

By the time he reached New York he was resentful, and he was poor, and he was bright—just the man to coldly undercut another man's work. In time he would redeem himself, but not yet. There was a flaw in Bell's work, and Edison accepted Orton's assignment to attack it.

Bell's design depended on sending the vibrations of the human voice into a microphone, to start the electric current that would run through the wire stretching from one telephone to the next. But to get a signal to travel more than a few hundred yards, you had to yell, and the signal often died or became too feeble to hear before it got more than a few miles away. Edison thought about it and saw there was a way to keep an electrical signal going as it traveled further through the phone wires. Before anyone even exhaled into the phone, he had a dedicated battery pump a strong, steady electric signal through the wires. When the speaker began to talk, his breath had only to modify the already robust battery signal, making it a little bit stronger or a little bit weaker. The result was that the speaker's voice didn't fade so quickly, and phone messages could be sent dozens of miles.

Orton was delighted, and paid off Edison with the equivalent of several million dollars in today's money. But Orton's delight didn't last long, for although Bell was meek, his new father-in-law was not. There were lawyers hired, leaks to the newspapers;

it's possible there were some quiet threats to Orton. Bell ended up keeping the main phone patents, although Western Union got some income from the improved microphone.

None of this mattered to Edison and his team. For Edison's stint as a patent-breaker had led him to think some more about the way Bell used the resistance in a wire to modify a moving electric current. Other devices, he realized, could use the same twist. And indeed, on October 30, 1878, J. Pierpont Morgan wrote to his Paris representative:

"I have been very much engaged for several days past on a matter which is likely to prove most important to us all. . . . Secrecy at the moment is so essential that I do not dare put it on paper. Subject is Edison's Electric light. . . ."

Edison liked gruffly pretending to his friends and to visiting newspapermen that he was just a simple man who had no interest in anything more than patching together a few practical devices. But that wasn't true. When someone's smart enough to duplicate or improve an important invention, as Edison had done with Bell's telephone, he's usually smart enough to wish to come up with important insights of his own. Edison had tried to read through Newton's writings as a youngster. He wanted to make an original contribution to this new world of electricity in which his technical skill had allowed him to get rich. An effective lightbulb would be a good start.

For decades researchers had dreamed of making a practical artificial light, but no one had come close to succeeding. Anyone who had watched a cast-iron stove knew that heated metal glowed first red, then orange, and finally it might even glow white. If a piece of metal could be connected to a battery and heated up that much, it would produce light. But how to make the glowing metal last long enough to be useful?

This is what no one had managed. The microworld was so little understood that it was hard to control how electric power jumped out when it was tapped. As early as 1872, the Russian Aleksandr Lodygin had placed two hundred electric lamps around the Admiralty Dockyards in St. Petersburg, but when he switched them on, they burned so powerfully that the metal filaments melted in just a few hours.

The lure of an electric light didn't go away, though, for the oil or gas lights that were the best alternative had problems of their own. Great groups of whales had been destroyed in the early 1800s to get a relatively clean oil for lamps. When that got too expensive, kerosene and other heavier oils were used, producing, however, smoke, smells, and—when the lamps were knocked over—fires. Natural gas was a little better, but it was expensive and hard to pipe for any distance, and users had to keep on adjusting their lamp burners to keep streams of soot from billowing out.

The first metal that Edison considered for his electric lights was platinum, since it has one of the highest melting points of any metal known. But it's also one of the most expensive metals known, and pretty soon he moved on to cheaper ones, at one point thinking he might succeed with heated nickel wires. This didn't burst into flames as much as his previous tries, but even when it just glowed, the light was too strong: "Owing to the enormous power of the light," Edison jotted in his notebook, ". . . suffered the pains of hell with my eye last night from 10:00 P.M. to 4:00 A.M. . . . Got to sleep with a big dose of morphine."

In time he managed to build the nickel-wire lamps without staring at them, but they still burned out too fast. A colleague recalls one of his first demonstrations, to Wall Street backers: "Today I can see these [nickel wire] lamps rising to a cherry red, like glowbugs, and hear Mr. Edison saying 'a little more juice,' and the lamps began to glow. . . . Then . . . there is an eruption and a puff; and the machine shop is in total darkness."

The first trick Edison used to keep the filaments from burning out was to stop any oxygen from getting to them. That meant surrounding them with little vacuums. He bought pumps that would pull air out of glass containers, and he hired a top glass blower, and he improved the pumps, and before too long, there in his rural New Jersey laboratory, his team had created small glass containers in a shape that reminded onlookers of tulip bulbs—our "lightbulbs"—that had less air inside than is found at the top of Mount Everest, or even several hundred miles higher above the Earth. By late 1879 he had small glass bulbs that held barely one-millionth as much air as the ordinary atmosphere.

They still didn't work. Any metal filament Edison put at the center of one of these bulbs got so hot that it would burn or melt or crack or—despite the low air pressure in the bulbs—just sizzle along to failure. He realized he had to try something other than metal.

For a while Edison put strips of charred paper between two electrodes to see how well they would glow, and he also tried fragments of cork, and then cotton threads. The cotton seemed especially promising, and for a long time he trumpeted that as his great success. But in time that too failed, and in exasperation he examined the paper fragments under his microscope, only to find that he couldn't magnify them enough to see the electrical sparks that he imagined running through them. All he had was the belief that any gushing electric particles would bump and slap along inside one of his filaments, hitting so hard that the wire or thread would get hot—just as the friction of rubbing your hands together quickly makes your palms heat up. He decided to search for a smoother filament.

"I believe," he told his workers, almost in exasperation, "that somewhere in God Almighty's workshop there is a vegetable growth with geometrically parallel fibers suitable to our use. *Look for it.*"

And this his team did. He had more money than any of the other inventors working on electricity—those nearly limitless funds from his New York backers—and more important, he had the most motivated workers. Edison knew that his drive came from having been poor, and he generally hired others like him: there were tough, itinerant technicians who'd done who knows what in the Civil War; there was a bright London Cockney, Samuel Insull, and many others. The team had developed expertise in wire filaments and air pumps; now they collected learned volumes on plant fibers. When hunting through books still didn't yield an answer, they started traveling: one worker to

Cuba, another to Brazil, a third to China and other points east. And there, in south-central Japan, they came across the Madake bamboo. It had a fiber far better for Edison's needs than platinum, nickel, or even the highly scorched cotton that had been the best till then.

When Edison's men connected strands of Madake bamboo to the wires from the battery metals and turned the battery on so that powerful charged electrons poured out, a faint glow came from the bamboo. When they slipped a glass bulb around the bamboo and pumped the air out of it, the bamboo strand got brighter, and would glow and glow and glow. The platinum bulbs in Russia had lasted twelve hours at best; efforts by Joseph Swan and others in England, around the same time as Edison's experiments, had reached a few dozen hours. But the Japanese bamboo, glowing away in its airtight bulb, as isolated as if it were in the vacuum of outer space, lasted for more than 1,500 hours.

To make his invention truly practical, Edison and his men had to create numerous related inventions. Their first impulse, as always, was to steal from other patents. But they were venturing into such fresh territory that it wasn't always possible simply to copy other people's work. The electric bulbs had to be easily fitted into sockets, for example, yet no one else had needed to do that, so the team came up with an original way of modifying the screw stoppers of kerosene cans (whence our screw-top bulbs today). They attached the vacuum bulbs so tightly to the screw that no air would seep in and make the glowing filament burn too fast.

Still more inventions were needed. They needed a system of automatically measuring the electricity that was used (so they could then bill for it), and there had to be improved ways to power the bulbs, and soon Edison and his team had so much new ground to cover that, without realizing it, they'd almost entirely stopped copying patents. A single telephone could be invented by a single individual. But Edison's network of power stations required dozens of synchronized developments in switches, fuses, power lines, underground insulators, and the like. Edison wasn't a cheat anymore. He was a creator.

This late-1870s surge of invention went far beyond the development of the telegraph a half-century before. The telegraph had seemed infinitely powerful; its series of innocuous clicks had changed business habits, financial markets, newsgathering, and political organizations around the globe. Transferring information faster shrank the globe, just as the lightbulb shrank the night.

But no matter how far a telegraph's signals traveled, the only thing "created" at the other end of a wire was simply a clicking sound. Victorian engineers had been able to make huge objects move, as with locomotives, and factory pistons, but that depended on big clanking steam engines. Now, in the final decades of the nineteenth century, they devised one way after another of pouring charged electric particles into new devices, and using that power to make them move in fresh, ingenious ways.

The most powerful of these creations was the electric motor. There had been small, toylike motors around for several decades, but, as with the telephone, Edison and his team—along with many others—made them a lot better.

To understand what goes on inside a motor, imagine a clock face with only a single long minute hand, pointing straight up to twelve o'clock. That minute hand wants to stay still, but someone has pressed a small electromagnet into the face of the clock, smoothly recessed into the surface right where the three o'clock mark would be.

When the electromagnet is switched on, the metal minute hand has no choice but to start turning clockwise, for it's pulled toward the beckoning magnet. If the magnet stayed on, the minute hand would stop at three o'clock, held quiveringly in position by the magnet's pull.

Instead, imagine that just before the minute hand hits three, some tormentor turns off the magnet and switches on a second electromagnet at the nine-o'clock position. The metal hand would whir past the three-o'clock position by sheer momentum, but then, instead of slowing to a halt, would start to feel the pull from the magnet at the nine-o'clock position.

If the trickery stopped there, the minute hand would reach nine o'clock and finally come to rest. But no, imagine that just before it reaches that destination, the nine-o'clock magnet is switched off. The metal minute hand skims past it, the three-o'clock magnet is quickly turned on, and the whole ridiculous spinning motion is repeated. The minute hand is like a greyhound racing after mock rabbits, perpetually kept out of reach.

That's an electric motor. (You can often hear this mechanism inside a running motor, for if there are two electromagnets, each whirring to a start and stop 110 times per second, there will be

220 separate whirs. That creates a hum not far from middle C.) To get power from it, you just grab on to the part that's spinning around. Going back to our imagined example, if you hung a thread from the minute hand of this peculiar clock, the electromagnets luring the minute hand along in a circle would only be strong enough to tug along a doll-sized wicker basket dangling from the thread. Scale up the device, though, as Edison and others did, so that giant electromagnets are providing the power at the three-o'clock and nine-o'clock positions in the motor, and the metal rod that's being tugged in a circle will be powerful enough to drag a ton or more of elevator straight up a shaft in a tall building.

This was crucial for skyscrapers. Strong metal beams were necessary too, but there would have been little enthusiasm for tall buildings if users had to climb up several dozen flights of stairs. With the electric elevator, no one had to. Land prices were high in central New York and Chicago, so it made sense to build vertically. Soon the skylines of those cities and then others became spiked with these tall, electrically navigated buildings. Electric charges that were billions of years old were now being manipulated to pull Victorian office workers up these narrow elevator shafts.

If electric motors were a bit smaller, with a rotating metal rod just a foot or so in length, the spinning wheels of electric streetcars could be built. This produced a momentous change, for more and more people no longer had to live within walking distance of their factories or offices. A small number of wealthy individuals who could afford horses and carriages already were able to live like that, and steam–powered locomotives had made some mass commuting possible. Now even more people could do it. Long–stretching suburbs burst into existence, growing along the new streetcar lines.

Electric motors did more. Streetcar companies that had in–stalled big power stations to pull their cars found that after seven P.M., when workers were at home, there was little use for their product. What could they do with the spare capacity? One solution was the invention of the modern amusement park. Electrically operated rollercoasters and brightly lit arcades ap–peared at the edges of cities throughout America and in parts of western Europe. By 1901, most of the largest U.S. cities had such amusement parks, powered and owned by local electric street–car companies.

There was a lot of mingling in the amusement parks, and it was often of a sort that the older generation didn't want. Before inexpensive travel to such parks, poor and immigrant children usually socialized in the neighborhoods where they lived. It was easy for parents and neighbors to keep an eye on them. But when the kids could go to these new parks and meet almost anyone, that control broke down. Sometimes there was fighting and arguments as different groups collided; often, though, there was courting, and furtive kissing, and increasingly, marriages that cut across traditional boundaries.

Industry changed, for the place where energy was made could be far from the place where it was used. The cable cars of

San Francisco had been among the first mobile devices to rely on this principle, since lifting a heavy iron motor up and down San Francisco's hills was too hard even for steam engines. Factories that were run by electricity could now use the same techniques. Workers didn't have to huddle around tools that sat close beside a steam engine or were powered by a single long pulley belt. It wasn't even necessary to have a steam engine and its heavy coal supplies on the premises. Just as with the San Francisco cable cars, power could be produced dozens or hundreds of miles away and then fed in. Industrial cities expanded, even where there were no waterfalls or coal.

Revolution was everywhere, even in the home. For the first time in history, glucose stored in human tissue was no longer the sole energy source available to power dreary domestic activities such as carrying, cleaning, and washing. Small electric motors took over many of those tasks.

This shifted relations that had seemed locked in since time immemorial. When servants are on their knees banging sodden washing, or scrubbing out blackened fireplaces, or trudging up and down stairs with slopping buckets, they seem so different from someone with free time for conversation or reading that it's easy to imagine that the servant doesn't "deserve" to vote. (In their exhaustion, servants can also feel that demanding the right to vote is too grand as well.) But with electric pumps and motors to run washing machines, and later with electric refrigerators and sewing machines and more, there was less menial labor and, with that, less subservience: votes for working-class men—and then the full heresy, votes for women—began to seem possible.

Edison should have been happy, for he and the R&D teams he led played important roles in almost all these technical inventions. Although he complained a bit when he got older, he loved gadgets and accepted most of these social changes. But he was still not satisfied. He continued to puzzle over the underlying science in a way that few of his fellow engineers did.

He was supposed to be the greatest electrician of his age, yet he didn't even know what was happening inside an electric wire. Most of the time, when journalists asked him to try to explain how these great inventions really worked, Edison would just laugh them off. He'd say that those matters were for the fancy professors to work out, that he would be long dead before that happened. Once, though, Edison did come across a hint. In 1883 he noticed that a black spot occasionally appeared on the inside of one or another bulb he was testing. This was odd, because the glass was always spotless when it was sealed around the filament. The dot couldn't have been a scratch (the filament never touched the glass), nor could it have been dust or soot (there was scarcely any air inside the bulb to carry dust).

Edison was puzzled by the dots. Was something flying out from the filament to create a dot? He wanted to explore further, but his assistants backed off. If it was a practical invention, they'd have worked any hours to help the Old Man. But small black dots? Edison tried to keep investigating on his own, but it was hard without support, and in a few months he gave up. "I was working on so many things at that time," he once said, many years later, "that I had no time to do anything more about it."

It was the mistake of a lifetime. In the next dozen years a few other researchers started looking at this and similar findings. The most persistent was a man just a few years younger than Edison, Joseph John Thomson, working in England in the 1880s and 1890s at the same Cambridge college where Newton had worked.

Thomson was not the most likely of experimental champions, for even his friends—who called him J.J.—winced at his difficulties in actually building the devices for experiments he could so easily plan. (He's easy to recognize in an official photograph of the Cavendish laboratory from that time: he smiles weakly from behind thick glasses, the only one with his tie askew.) But he managed to build enlarged versions of Edison's lightbulbs, and used magnets to invisibly reach inside and "steer" whatever was flying up from the filaments. Then he weighed the flying particles.

That's how he discovered the electron. Atoms were not solid little balls. Rather, parts of them could be torn off. The torn-off bits could bounce and skid forward, like smaller balls, within any open channel that lay ahead of them.

It's those torn-off bits—electrons—that, as we saw, roll forward inside a wire, creating an electric current. That's it.

The quietly bumbling J.J. had managed to explain what Newton and so many others had only guessed at. It seemed so easy! The world is made of powerful electric charges, which are normally hidden, but which we can scrape loose. What pours out of the metals in batteries, it seemed to J.J. and his colleagues, were just miniature flotillas of those electrons, released after untold aeons of being locked inside.

When those escaping electrons bumped and crashed inside the filament of a lightbulb, their collisions made the filament hot enough to glow. Even the black dot that Edison had seen was simply created by electrons that had erupted from the filaments in his lightbulbs, their accumulated impact etching into the glass.

The century-long quest to see what was happening inside an electric wire seemed over. J.J. Thomson, rather than Edison, got the Nobel Prize, and was heralded as the man who explained how Victorian electricity really worked.

But there was one giant flaw.

Was it really true that electrical apparatuses merely worked by electrons rolling along inside? If that was so, then when someone in New York spoke on the phone to someone in Boston, they'd be pushing electrons from the telephone in New York through a metal cable until the electrons popped out of the receiving phone in Boston. But that didn't make sense. If the New Yorker talked for a long time, without letting the other person interrupt, would huge black dots begin to accumulate in Boston, as a great mass of electrons started piling up? That never happened. Somehow the explanation was incomplete.

There had to be something else in the universe—some invisible force that controlled the way electrons moved; a force that could perform the seeming miracle of making these electrons travel *without* piling up at the far end. But what could it be?

Edison was convinced this invisible force existed, and he had even once tried to contact it. In absolute privacy he'd built a small pendulum, attached a wire from the pendulum to his forehead, and then tried using the sheer power of thought to move the pendulum. Nothing happened and, half embarrassed, half puzzled, he'd put the experiment away, accepting that he was not going to be the one to reveal any such unseen power.

In fact there was a whole group of investigators, trying to identify and understand these further powers of electricity. They'd been at it for many years, but their work was so theoretical that it had been ignored by most of the practical inventors of the 1800s. These investigators believed that all of humanity was surrounded by a powerful network of mysterious force fields. According to them, people had been walking through these fields for many thousands of years—in Mesopotamia and Egypt, in China and the Andes—but because the fields were invisible, no one had ever noticed they were there. The only hints of their existence were "mistakes" of nature, such as the sparks caused by static electricity or the flash of lightning.

Edison vaguely knew of the researchers who held these beliefs, and he also knew that something important to them had taken place in a strange engineering project deep under the Atlantic Ocean in the 1850s, when Edison was still a child. He also knew that even before that engineering project, there had been a great English scientist, Michael Faraday, who'd predicted the existence of these invisible force fields.

As a young man, Edison had tried reading several of Faraday's works, but now that he was so busy with lightbulbs and generators and electric motors—now that he had a vast work-force to supervise, and a great personal fortune to invest—it was much harder to find the time for such difficult reading. He could only occasionally try to imagine what powerful new machines that invisible force might produce if it were ever mastered.

Edison and Bell and the other practical Victorian inventors had thought they'd reached the deepest core of things, when they let out the ancient power of electrons. But they'd only scratched the surface. Underneath, there was something more.

PART II

WAVES

*T*he metals that escaped from ancient exploding stars had a further great power when they fell to Earth. Around each of the landed electrons hidden within them, an invisible force field stretched outward.

Under normal circumstances that field was impossible to detect, but the invisible field was often stronger than usual when it emerged within iron. Much of the ancient iron had seeped deep into the planet's depths, and as the planet spun, this iron spun as well.

From that spinning, a planet-embracing magnetic force field was created.

It spread up from the ground, and although it was unobserved for almost all of human existence, in time the classical Chinese began to detect it. They used some of its power to orient delicate swiveling compass needles for navigation.

Most of the world's residents were too poor, and most thinkers were too dogmatic, to go any further with the hint provided by these pervasive magnetic fields. But then, in the thirteenth century of the Mohammedan era, and the twenty-fourth century since the mission of the Buddha, and the fifty-sixth century of the Jewish calendar—in other words, in the nineteenth century of the Christian era—that fundamentally changed.

4

Faraday's God

LONDON, 1831

Michael Faraday, the man who did the most to uncover these invisible force fields, was a curly-haired, working-class Londoner who was born in 1791, more than a century before the electron was discovered. He was exuberant as a teenager, teasing his friends with wordplay and games as they raced down the London streets, replying once to a letter from his friend Benjamin Abbott:

—no—no—no—no—none—right—no Philosophy is not dead yet—no—no—O no—he knows it—thank you—'tis impossible—Bravo.

In the above lines, dear Abbott, you have full and explicit answers to the first page of yours dated Sept 28.

Even when he wangled his way into a job as laboratory assistant at the august Royal Institution in 1813, he kept the same humor, on one memorable evening sneaking Abbott in so they could sample the nitrous oxide—laughing gas—which the director of the institution had put aside for one of his experiments.

But Faraday also had a serious side, and was drawn to the same electrical puzzles that Joseph Henry sought out. How could a coiled copper wire act as a magnet and pull lumps of metal toward it? There was only empty space between the wire and the metal. Nothing in conventional science made sense of this. Hold such a live electromagnet above a nail, and the nail will fly upward. Yet the gravity of uncountable trillions of tons of rock and magma—the entire mass of the Earth—is tugging down on the nail.

What pull could a magnetized coil of copper send out that would overcome that vast force?

Faraday was fascinated—it's what he'd dreamed of unveiling—but for years he was barely allowed to work on such problems. Rumors were being spread behind his back that this onetime slum child, this offspring of a mere blacksmith, was not really capable of serious research. But by 1829 the Royal Institution director who'd slurred him the most had conveniently died. Faraday gave his most sincere compliments to the director's widow, immediately dumped the assignments the director had given him, and cleared his work schedule as much as possible. He couldn't put the invisible drawing power of magnets out of his mind. He had to know how it worked.

In this investigation, Faraday had one great advantage over his rivals in England and on the Continent. They had all been trained in the advanced mathematics that Sir Isaac Newton had developed in the seventeenth century. Newton was famous for the image of a cold, clockwork universe, where planets rolled along like giant, separate billiard balls. There was no place in that universe, they were now taught, for invisible forces to fill up the space between solid objects, holding the solar system or the universe together; there were no invisible cobwebs pulling across the sky. Gravity existed, but it somehow leaped from object to object. In this view, it did not permeate the voids in between.

This meant that when Faraday's contemporaries tried to understand the links between magnetism and electricity, they assumed it would have to be a force that leaped across a gap, without really existing in the space within the gap. Their universe was basically empty. When forces operated, it would have to be, they believed, through the cold, distance-leaping process that Newton had labeled "Action at a Distance."

Faraday respected Newton, but he'd been supporting himself from the age of twelve, for years as an apprentice bookbinder. He'd learned to think for himself: if he'd gone along with what everyone assumed, he would probably still be at the binding shop. Also, in his years as an apprentice, he'd skipped learning much mathematics beyond elementary arithmetic. This had the further advantage that he'd never been seduced by the beauty of Newton's Bach-like equations. But even if he hadn't been poor, and even if he had learned calculus, there was a further reason Faraday wouldn't have been convinced that space was empty.

Faraday's family had been devout members of the gentle religious minority called the Sandemanians, a Quaker-like group with a near-literal belief in the Bible. Even as Faraday moved into the Royal Institution, he remained a devoted member; his wife was a Sandemanian, and his closest friends were as well.

From his religion, Faraday was convinced that space was not empty, but that a divine presence was everywhere. He was used to being ridiculed for such beliefs ("I am of a very small & despised sect of christians," he once sighed), and he had learned to keep his views private. But religion dominated all his thoughts, and once, on a small boat in Switzerland, Faraday saw what he took as a proof of his beliefs.

It was an ordinary rainbow at the base of a waterfall. But a strong wind was blowing, and the gusts often threw the spray so far to the side that the rainbow disappeared. When that happened, Faraday motioned for the guides he was with to wait. Every time, the wind blew the spray in the opposite direction, and the rainbow reappeared.

"I remained motionless," Faraday wrote, "whilst the gusts and clouds of spray swept . . . across its place and were dashed against the rock." It was, he felt, as if the rainbow were always waiting, even though it could only sometimes be seen. That was what he believed of science. Even when space seemed empty, something was there.

Now, looking for a further link between electric currents and magnets, he knew to concentrate on the one thing everyone else had missed: the apparent "emptiness," the voids between the different objects in his lab. He used a simple clue as the way in.

A common parlor trick of the time was to sprinkle iron filings near a magnet and then watch as they formed curves that stretched from one tip of the magnet to the other. To Faraday this wasn't just a trick to amuse children. For where did the sudden arcs really come from? They were a sign, just like the rainbow, of the invisible matrix for which he'd been searching.

Through 1830 and increasingly in 1831, Faraday began circling his quarry. He became fascinated by the way forces seemed able to jump from one domain to another. When clouds of gas vapor ignited, for example, it was common to say that a ball of flame instantly appeared. Faraday didn't believe that. When he stared more closely, he felt he could see the flame propel itself very quickly from one part of the vapor to another. When he went to the seashore in Hastings, a long day's coach ride from London, his wife found him kneeling on the beach to examine the ripples in the sand, pondering how they spread. It was something that Edison would never have had the time to do.

By the spring of 1831 he was close, but still hadn't found what he wanted. He was thirty-nine years old, and in many of the years the Royal Institution's director had kept him at bay, he hadn't come up with any significant discoveries. Had his critics been right in suggesting that he wasn't a first-rate thinker after all? Faraday cut back on his lecturing and began arriving even earlier at his lab. Occasionally his two nieces would come downstairs to visit, but they knew they'd have to spend most of their time sitting quietly in the corner, cutting out paper shapes or playing with their dolls while Uncle Michael worked. Months passed, colleagues wondered what was happening, and then—in work that came to a head in those pressure-soaked weeks just before a fortieth birthday—one of his oldest friends, Richard Phillips, received a quick note:

Sept. 23, 1831

My Dear Phillips,

. . . I am busy just now again on Electro-Magnetism, and think I have got hold of a good thing, but can't say; it may be a weed instead of a fish that after all my labour I may at last pull up. . . .

It wasn't a weed, and after a few more days' work he had the decisive result. By October he was able to present it in an immensely simple form. He merely held a child's small bar magnet in one hand and a coiled wire in the other. He pushed the magnet toward the coil of wire, and an electric current started up in the wire. He held the magnet still; the current stopped. He moved the magnet again. The current started up again. So long as he moved the magnet in the vicinity of the wire, he created an electric current.

No one, ever, had understood this before. He'd created a force field! Something was traveling from the magnet into the wire. But that couldn't happen if the space between them was empty. There in his cool basement laboratory at the Royal Institute, with the horse-drawn traffic of Regency London pounding outside, Faraday had shown that electricity wasn't some hissing liquid that could only be funneled along inside a wire. Rather, it could be brought into existence by an invisible force that spread from a moving magnet and stretched across empty space.

Faraday had opened the door to something greater than

anyone had imagined. If he was right, then whenever his nieces playfully tugged a toy magnet, they were also tugging an invisible force field that spread out from the metal of that moving magnet. As best Faraday could estimate, the force field stretched on forever. If he and his nieces were inside a building, part of the force field would stretch outward through an open window or perhaps even through the wall, absolutely invisible, and continue to the moon or possibly beyond.

It got stranger. Faraday's basement experiments suggested that our world was filled with untold millions of these invisible, flying force fields. There were hundreds of ships in the London ports and thousands of carriages on the London streets, and whenever any mariner or coachman's magnetic compass needle moved, yet more of these invisible fields would be loosed. When Faraday looked out at Regency London, the sky wasn't empty above him. It was arcing with these powerful, invisible presences.

"The book of nature, which we have to read," he'd once written, "is written by the finger of God." He was right, and now he had shown that God was a flamboyant, hallucinatory Titian, blasting His universe with hitherto unseen, vivid streaks.

Faraday's insights are at the heart of modern technology, and even—as we'll see—would eventually answer the question about why electrons don't pile up at the end of a long phone line. But although he now held a distinguished position at the Royal Institution, he hadn't outlived his past. Most of his English colleagues believed that he was only a clever tinkerer. They knew about his embarrassing lack of formal education; they'd seen that he couldn't express his insights in the advanced mathematics they so easily used. To all but a few of them, his excited theories of invisible force fields seemed entirely unfounded, and so his ideas were politely set aside.

Faraday made many more discoveries; he had audiences with prime ministers; he became greatly respected for his popular lectures. At one point a brilliant young woman became fascinated by what his electricity findings might suggest for her own research. This is one of the great "might have beens" of history, for the woman was the late Lord Byron's daughter, Ada, Countess of Lovelace, and she'd been working on early notions of what we'd now call computer programming. No technology of the time could entirely build what she envisaged, but who knows what Faraday might have come up with? He seemed entranced by her, but soon stepped back, probably to avoid jeopardizing his own marriage.

He still didn't give up, though. When he'd been criticized for his religion, he had turned to the Bible for consolation. Now, criticized by the great majority of researchers for his idea about force fields, he turned back to Isaac Newton. Although Newton was said to have had a different view about empty space, maybe that wasn't quite true. Newton was the greatest thinker science had ever produced. Even a hint in his writings that Faraday might be right would be comforting.

On the surface it would seem impossible, but in fact Newton had once revealed doubts about his public vision, in a brief relaxed moment in his old age. To an inquisitive young Cambridge theologian, Richard Bentley, Newton wrote, in 1693, that perhaps the universe wasn't so empty. Perhaps, on the contrary, there really were forces, such as gravity, that sent out tendrils crisscrossing what seemed like empty space. The idea "that . . . one body may act upon another at a distance through a vacuum without the mediation of anything else," Newton daringly wrote, ". . . is to me so great an absurdity that I believe no man who has . . . any competent faculty of thinking, can ever fall into it."

Bentley had been excited and had written to find out more about what the great scientist meant. But Newton drew back. These were just an old man's musings, and nothing further was to be said. To have gone further would have been dangerous, for this was an era when religious heretics were still burned at the stake. The authorities might misinterpret his belief that space was not empty and assume he didn't believe God's power was great enough to cross empty space, they might begin to investigate his private religious writings, which definitely were full of heresy. The correspondence stopped, and his brief hint of self-doubt in that letter to Bentley was soon forgotten.

But now, more than 140 years later, as Faraday was hunting for confirmation that he hadn't been entirely wrong, he found Newton's very old letter to Bentley. With that, Faraday realized he wasn't alone. Newton had been there before.

Beyond that, Faraday couldn't go. In old age, his memory already fading, he wrote to a young friend, the gifted Scottish physicist James Clerk Maxwell:

Royal Institution
November 13, 1857

My Dear Sir,

   . . . There is one thing I would be glad to ask you. When a mathematician engaged in investigating physical actions and results has arrived at his own conclusions, may they not be expressed in common language as fully, clearly, and definitely as in mathematical formulae? If so, would it not be a great boon to such as we to express them so—translating them out of their hieroglyphics that we almost might work upon them by experiment. . . .

Maxwell wrote back diligently, but Faraday remained left behind. His weakness in mathematics had helped him start a major inquiry, but he was never going to be able to lead its further development in his lifetime.

Still, Faraday consoled himself by taking a long view. He was convinced that someday there would be practical inventions that depended on what he'd seen. When that finally occurred, even the critics who dismissed him would have to accept that his hunches were true.

What he didn't realize, as the decades went on and he was reaching sixty, was that he would live long enough to see it happen. A giant engineering adventure was soon to take place, deep under the sea. When it was completed, there would be tantalizing further evidence that everything he'd imagined about invisible force fields really was true.

5

Atlantic Storms

HMS *AGAMEMNON*, 1858, AND SCOTLAND, 1861

The undersea adventure that finally resolved Faraday's hunches began with one Cyrus West Field, who was musing, on a cool January afternoon in 1854, before an ornate globe in the library of his New York brownstone. The telegraph to which Morse had put his name was ten years old; Thomas Edison was just a seven-year-old boy in Michigan. Field had made a fortune in business, but although he was supposedly retired, he was still only in his mid-thirties, and he didn't especially like being retired. He'd tried explorations to South America, but those were just guided tours for rich men, and the incident when he'd brought both a jaguar and an apparently nonviolent Amazonian teenager back to New York City didn't bear remembering. He'd also tried spending time in the social world of high-caste New York, of the sort the novelist Edith Wharton would later write about, but the obsessions with tea and gossip had convinced him that the Amazon was not so bad after all.

Now, in his study, he noticed something . . . annoying . . . about the globe. On one side there was England and its empire, the great source of proper white culture, while far away, isolated by a vast ocean, there was America. Why should these two natural allies be so crudely separated? The only way to send messages between them was by ship, and it could take weeks to make a round trip between the two nation's capitals.

In the past, that had produced great misunderstandings. The great battle of New Orleans in January 1815 had lost some of its grandeur when the British and American forces involved were informed that the war they were ostensibly fighting had ended several weeks before. The news had taken that long to reach deep into the South.

By the 1850s those problems were being remedied when it came to land communications. Most great cities with an over-land route between them were linked by telegraph: Washington and Baltimore, Paris and Berlin.

A few telegraph lines had even been dipped for a few miles underwater, as with the line that had been operating for several years now across the English Channel. But the great challenge—the chief remaining adventure, Field realized—was the colossal stretch of the North Atlantic.

Its surface was rolling and treacherous for thousands of miles, but underneath, far in the depths, was it not possible that the advanced electrical works of man, lowered cautiously to these regions, might survive, untouched, for decades on end? Many individuals may have had this vision—in just a few years the young Frenchman Jules Verne would come up with his Captain Nemo stories, about a brilliant sea captain and an advanced submarine that spends months at a time in these depths. But Field had the money to make this vision of connecting the two continents come true.

He would build a cable, he decided, a giant ocean-spanning cable, and he would link the two great empires, and he would produce universal brotherhood, or at least great profit, and in doing so he would get to leave this cursed brownstone and the rituals of stuffy New York life and instead travel to the world's financial center, London, and work with engineers, and sailors, and salty sea captains.

It turned out to be agonizingly harder than he thought. Over the next fifteen years, Field made dozens of crossings of the Atlantic, vomiting almost every time; there were cables that cost millions of pounds sterling that snapped far from shore; there were high storms, and swindlers, and parliamentary inquests, and an attack by a whale, and an embarrassing occasion when New York's City Hall nearly burned down. But even if Field had known all that, he probably still would have gone ahead. It was better than being bored—and what did financial loss, whale attacks, and seasickness mean against the hope of glory?

Most of all, in his heart, Field believed the actual technical operation was going to be easy. To him, the cable was like a hose or a narrow tunnel. Electric currents were some sort of crackling, hissing stuff that batteries mysteriously created. He'd just pour that into the hose. If the signal that emerged at the far end of his cable was too weak, well, he'd simply pour in more.

Ultimately Field realized that there was far more to what

happened, but for now, when he arrived in London in 1854, resplendent in the best outfit his New York tailors could prepare, he received the eager welcome that any confident American with a great sum to invest is accorded. Everyone thought he had a fine idea, a cracking idea, and those who'd worked with telegraphs assured Field that they'd done the tests to prove his ideas, or at least they were about to, but in any event the accuracy was in the bag, and if Mr. Field wanted a trustworthy partner, he couldn't find a better soul.

Field was polite enough, but he had made his way up in the New York textile and paper trade—where trust in human intentions was rewarded with extinction—and, like the majority of Americans arriving in London with a great deal of money to invest, he was quite aware that most of the people he met were trying to take him for a fool. He didn't commit himself, and discreetly sought out the leading electricity theorist in the country, just to be sure his views were right.

This was the Scottish scientist William Thomson (no relation to J. J. Thomson). One visitor who came to meet Thomson around this time was ready for a white-bearded ancient, but instead found an energetic young man in his thirties who raced up the stairs two at a time. Thomson was a fit sailor and had been a champion rower and swimmer at Cambridge. He'd also received the highest scores in his year on the final examination in mathematics—helped, most conveniently, by the fact that at least one of the questions was from his own prize-winning published papers.

Cyrus Field knew of him because Thomson had been making a study of the few undersea cables that were already in operation. What he had found was not appropriate for wide dissemination—it could prove disquieting for Imperial morale—but a series of disturbing flaws had been noticed in each one of Britain's cable systems.

Although foul play by one of the opposing Great Powers could have been considered if there was only one incident, the mysterious flaw also cropped up in several Mediterranean cables, and even in the London–Brussels line. Signals that were sent in sharp and clear—a single brief flash of electricity—were no longer sharp and clear when they emerged. Instead, they were muddy and blurred.

With a short cable the flaw was just about tolerable, but with longer distances it meant anyone who used the cables—and the Admiralty increasingly depended on them—had to send and re-send important messages. On a *very* long cable, such as the one Field proposed to lay beneath the Atlantic, it would—unless understood, and fixed—make clear messaging impossible.

No one had observed this problem in ordinary aboveground telegraph lines, however long. But why not? Something unique was happening in the undersea cables—and Thomson thought he understood what it was.

Thomson was one of the few thinkers then taking Faraday's vision seriously. He believed, with Faraday, that the surface of reality was misleading, that underneath the sparks and cracklings of a moving electric current there really was a deeper power, an invisible force field, and that was what pushed the current forward. The "sparks" (the electrons) that tumbled along inside a wire didn't move by their own power, but were transported as though by an invisible flying carpet. For Thomson and Faraday, the carpet was the invisible force field.

In the years since Faraday's first conception, Thomson had taken the idea further. This invisible field is what would emerge from a battery, he believed, and that was more important than any sparks. The field would travel partly within but also along-side any wire stretching ahead of it. It would take up position along the whole length of the wire, very quickly, and then it would reach in and pull forward any charged particles—any electrons—it found near its path. Thomson imagined the field as almost a living thing, constantly writhing and twisting, as it carried this incredible pulling power.

That's what especially worried him, for Thomson knew that the Atlantic cable Cyrus Field was planning to build would be in three layers. Each was to be as thin as possible to save weight. There would be a thin copper strand at the center, a thin layer of rubbery insulation around that, and finally a casing of iron wrapped around the whole thing, so that the cable wouldn't be ripped open as it was dragged and bounced along the deep ocean floor. That made sense in terms of Cyrus Field's view, but was terrible to Thomson. For when a telegraph operator tapped his key, the field would start racing forward alongside the thousands of miles of copper, yet it would also writhe sideways across all that length of rubbery insulation, and some of it would try to pull on the electric charges hidden in the thousands of miles of iron casing, and a final part of it would even get dispersed in the millions of tons of cold seawater outside. It would be spread wide, its energy dispersed.

This explained the delays the Admiralty and other investigators were worried about. When the human finger clicking the telegraph key lifted up between each signal, the field that had taken position along the thousands of miles of cable during the previous click would have to disperse before the next signal could go through. It would have to collapse back, down from the water to the iron, and from the iron across the insulation into the copper, and then from the copper it would disappear. If the finger tapped too soon, the new field that came charging in would collide with the old one still twisting around between copper and iron and sea. No wonder the signals in the few undersea cables that existed grew blurred and wobbly. (This didn't happen in ordinary land telegraphs, where the wires were held up on poles, because those wires could have thick insulation. Nor was there an iron casing to lure the sinuous force field outward. Any part of the field that did escape just chased off through the air harmlessly.)

Cyrus Field was a polite man, but these must have seemed the ravings of a lunatic. Thomson saw the field as a genie—a howling wind—struggling to get out. To get the project to work the whole structure and operation of the cable would have to be changed. For a start, the rubbery insulation would have to be much thicker to keep it in. But there were no genies in Cyrus Field's world, no force fields. He'd already paid good money for an advance order for cable with thin insulation. He wasn't going to change that now.

————

Field gave his compliments to Thomson and selected a more practical man as chief electrical engineer for the project. This was Edward Whitehouse, who didn't believe in preposterous, invisible flying force fields. For him, electric charges just shot out from the metals inside a battery and poured down the wire. There was no need for lurking force fields to fly alongside and speed them ahead.

Even better, Whitehouse was able to help Field in another, somewhat delicate matter. For Field was not quite as rich as everyone believed. He'd made a fortune, but not all of his subsequent investments had done well, and unless he got funding for his cable project, and quickly, he'd have to return to New York in disgrace. He couldn't fund the project on his own; he couldn't fund a tenth of it, or even a hundredth. No sign of hesitancy or uncertainty could be made public until his investors were in hand.

Whitehouse was the ideal man to guarantee that no questions were raised. He used threats to make sure that the handful of young scientists who supported Thomson's hypothesis stayed quiet or recanted; he even, humiliatingly, brought the elderly Michael Faraday to a public meeting to back Field's project.

Faraday had been suffering increasing bursts of confusion over the years—possibly due to mercury fumes coming from the floorboards of his laboratory, which can affect the brain if inhaled over long periods—and Whitehouse had carefully prepared him. Faraday's great strength was experimentation, everyone knew that, and Whitehouse seems to have misled Faraday about experimental evidence suggesting there were flaws in Thomson's calculations.

Under Whitehouse's pressure, Faraday gave an ambiguous statement suggesting that Thomson hadn't been entirely correct. In a move that would do twenty-first-century biotech venture capitalists proud, Whitehouse and Field spun that statement to give the impression that they had Faraday's full support for their narrow cable with only a thin layer of insulation. Respected public figures such as W. M. Thackeray started buying shares. Soon Field had enough cash for the cable factories to run full time, and for negotiations with the Admiralty and the American navy to proceed.

Thomson realized his idol Faraday was being manipulated, but what could he do? He was convinced he was right, but recognized that all he had was still only a theory. The cable project had too much momentum, and he was now ignored. Whitehouse blocked his letters to Field. When Thomson proposed an improved transmitter for the cable, Whitehouse ridiculed it and refused any company funds to build it.

On June 10, 1858, the British battleship *Agamemnon* and the U.S. Navy's *Niagara* sailed from Plymouth in England, ready to lay the cable. (There had been an earlier try in 1857, but the cable had snapped in water too deep to dredge it up.) Taking into account the insulation and the metal sheath, the cable weighed so much—almost a ton per mile—that no single ship could hold enough. So the two vessels headed to a central rendezvous, where they would splice together the two halves and then sail in opposite directions—one to Ireland, one to Newfoundland—playing the cable out between them.

Field was on board the *Agamemnon*, and so was a tough chief

engineer, Charles Bright, but when it came time to sail, Whitehouse didn't show up. The British navy had lost many big ships
in Atlantic storms over the years, and Whitehouse knew that it
never hurt to be too careful. He also wanted to avoid William
Thomson, who had talked his way on board.

Thomson knew that Field didn't believe his untested deduc-
tions, but he wanted the chance to show that his and Faraday's
ideas really were true. Also, he liked Cyrus Field and was going
to help him succeed, whether he wanted it or not. Whitehouse
had happily bullied Thomson in their correspondence, but he
knew of Thomson's intellectual reputation and was not about to
dare having a technical argument with him in front of Field.

Anyone who joked about Whitehouse's unnecessary caution
would have changed his view when the storm of late June 1858
broke. It was one of the worst recorded in the North Atlantic
in the nineteenth century. The *Agamemnon*'s captain, George
Preedy, seems not to have worried too much when his ship
tilted sideways so far that its masts dragged in the bursting
waves, but when its coal supplies started flying *up* through the
splintering deck, and then the tons of cable started following
them, he did confide to a London *Times* correspondent on board
that it was not, perhaps, ideal weather for laying a telegraph
line.

The battered ships met in mid-ocean, but there were more
storms, and repeated snapping of the cable, not to mention the
attacks by a whale, and a layover in Ireland while the cable was
repaired and fresh coal was taken aboard. Nevertheless, Captain

Preedy was damned if he was going to pull his ship out if the Yanks on *Niagara* were going ahead, and the U.S. captain of course reciprocated his views, which their respective seamen had no doubt emphasized to one another when they'd met for polite, articulate discussion at various waterfront pubs during the Irish break. And then, late in July 1858, the weather cleared, and for more than a week the Atlantic was as calm as an inland pond, the *Niagara*'s cable was pulled ashore in Newfoundland, and the *Agamemnon*'s dragged up to Valentia Bay, in Ireland.

As the news of the August 5 landing came out, the newspapers went wild. This was harmony within the dominant Anglo-Saxon breed, this was technology conquering the planet, and suitable celebrations were in order. In Britain, church bells were decorously rung, knighthoods were granted (including one to chief engineer Charles Bright, only twenty-six years old), and a formal congratulation from the Queen was printed in *The Times*. In America there were cannon barrages, preachers taking credit for the achievement, and torchlit parades (hence the unfortunate incident in which a lit torch met the highly combustible cupola atop New York's City Hall).

Field was a hero, and Whitehouse—who quickly took up position in the telegraph hut in Ireland, kicking out Thomson, who'd started connecting the wires before he arrived—was now going to be his prophet. For there were dozens of messages stacked up waiting to be telegraphed, and although the first few had been transmitted with no difficulty by Thomson, Whitehouse was determined to get credit for the rest.

He soon came to the greatest one—the formal congratulation from Queen Victoria to President Buchanan. The Newfoundland telegraph transcribers were ready; special correspondents from U.S. newspapers were standing by as well.

But something was wrong. The first telegrams had been transmitted without flaw, but now it seemed as if something inside the cable was changing. The Queen's telegram was only ninety-nine words, and a skilled operator could have tapped it out in a few minutes. Yet hours passed, there were hushed reports of members of Field's staff hurrying grimly in and out of the telegraph huts, and only well into the next day, after sixteen and a half exhausting hours of transmission, did the full text make it through.

The telegraphers' efforts went downhill from there. It took more than thirty hours of struggle with sending and resending to transmit Buchanan's equally brief telegram back to the Queen. Surviving documents show that the cable was increasingly taken up with messages of "Send more slowly," or "Repeat," or, most plaintive of all, long repeats of "What???" Newspapers became suspicious, the grand complimentary speeches turned to sarcasm, and soon Field's representatives no longer let themselves be seen in public.

And the cause of it all, increasingly frantic inside the Ireland transmission hut, was Edward O. Whitehouse. He was truly out of his element now, for everything Thomson had talked about was now coming true. Whitehouse's operators tapped in sharp signals, but they got blurry and spread out, becoming impossible to recognize by the time they'd crossed the Atlantic.

Whitehouse did what any bully might do when he's been caught in a bluff. He panicked. He had installed big backup generators at great expense in the Ireland hut—devices five feet high, like giant batteries, that could blast into the cable great amounts—well, of what? Whitehouse thought what came out of a battery was merely like the sparks from a welder's torch, only smaller; in later terminology, that a battery's job was just to launch fleets of powerful charged electrons into the wire, which then tumbled forward all the rest of the way on their own. Thomson didn't believe that. In his view, what shot out from a battery was a wild, powerful force field. Feeding more battery power into the undersea cable simply meant you were feeding in more of this howling field. That's why it was imperative, he'd explained, never to use the big batteries.

Whitehouse ignored him. He was a tough, practical man, who had made his own way up in English society. Strength had brought him this far, and strength would get him the rest of the way. There was nothing mysterious now, either. The signal wasn't getting through, so, he felt, he just needed to force in more of the tiny particles that were pouring out from the metals of his battery. In the time-honored English tradition of speaking louder to a foreigner—in the hope he'll understand you better if you shout—he told his underlings, now in these last weeks of August 1858, to hook up the five-foot coils and let them rip.

It was an astoundingly bad choice. Many people have noticed the peculiar warning on televisions and computers: BE CAUTIOUS IN OPENING DEVICE EVEN WHEN UNPLUGGED. Why should it matter, once the plug is out? The reason is that inside a

TV or computer, electric charges can build up on various metal surfaces. When the sweaty probing finger of an amateur repair-person pokes around and happens to place itself *between* two such surfaces, the electric charges on either side of the finger—waiting on those previously separated metal surfaces—suddenly have an easy path to travel across.

The inside of a small laptop would likely give you only a slight shock. But with the powerful batteries Whitehouse had hooked up, the inside of the Atlantic cable was now suffering contortions that made it much, much worse. He was pouring in a force field hundreds of times more powerful than Thomson had ever imagined deploying. Some of the force field reached into the central copper wire and pressed against the electric charges there, but much of it easily twisted sideways through the thin, rubbery insulation, and ended up using its energy to stir up electric currents in the cold iron outer casing. But re-member how the filament in Edison's lightbulb would heat up when a lot of electric charge was pushing through it, bumping and skidding against the main body of the metal atoms in the filament. The same thing happened here under the Atlantic. The central copper wire heated up, and the outer iron casing heated up, and the rubber in between was caught.

Each time Whitehouse commanded his operators to send yet another telegram, they had to make hundreds of Morse code keystrokes. This meant they were sending hundreds of surges of the invisible force field down the cable. The copper core and the iron sheath quickly got hot as this leaping field made strong currents start up in both of them. Eventually the rubbery insulation, sandwiched in between, didn't just get a little bit warmer. It began to melt. At each point where that happened, the charged particles inside the cable no longer had to travel the full, two-thousand-mile path forward. They could take a short-cut instead, simply traveling the now-unshielded single inch sideways from copper to iron. It was literally a "shorter" circuit—whence the common expression "short circuit."

The more Whitehouse tried to fix things—the more franti-cally he revved up his huge batteries—the worse the cable got. After a few days Whitehouse junked his own equipment, and—surreptitiously, swearing his operatives to secrecy—went back to Thomson's original, far more gentle transmitters and receivers. But the damage had already been done. Too few of the crackling electrical particles were able to travel undiverted through the long central copper wire; more and more took the nice conve-nient short circuits sideways to the iron insulation so nearby. Once there, they would fizzle off into the ocean, heating up the sea water slightly and never getting anywhere near the receiv-ing station, thousands of miles away.

By September only fragments of words were getting across, and by October 20 so much insulation had melted that no sig-nal at all could get through. The cable went dead and was never used again. Whitehouse was fired, and Thomson took over. The desperate directors of Field's company agreed to do whatever Thomson said. Did Thomson have any suggestions for them?

He certainly did. A new cable would have to be laid, he ex-plained, with much thicker rubber insulation, to try to keep the jumping field from getting across. Once it was connected, only very *gentle* battery pressure could be applied. Thomson didn't know exactly what was inside the copper core—this was still decades before electrons were discovered—but he did know that whatever carried the electric current in there was almost ineffably light in weight, so much so that the finest jeweler's balance scales couldn't detect it. That slight weight was all that he needed the force field from the battery to move. He was going to give Cyrus Field a whispering genie, not a bellowing one.

Cyrus Field still seems to have had some doubts about the theory, but he recognized he had no alternative. If he gave up, the project was over. Nor could he repeat the brute-force approach with which Whitehouse had failed. He'd have to gamble that Faraday and Thomson were right, and that invisible force fields that could move electric charges really did exist. It would, finally, be a detailed test of Faraday's extraordinary predictions.

There were more struggles and more delays, and in 1865 an improved cable snapped, heartbreakingly, two-thirds of the way across the ocean, in water so deep that it was impossible to dredge it up and fix it. But in 1866 yet another cable was laid, from the world's largest ship, the *Great Eastern*, and this one made it. From the time it was pulled up on shore, the new cable clicked along just fine, almost nonstop, for year after profitable year thereafter. Faraday was very old now, and very ill, but it seems that the news was delivered to him, possibly by young Thomson himself.

Thomson was proud, and Cyrus Field was rich. We know now that electrons aren't constantly geysering out of the sockets in your house. There's just a force field waiting at the sockets, led all the way to your house from a distant power station. When you plug something into a wall socket and turn it on, that guided field flies into your home, takes up position inside any computer or lightbulb, and simply yanks on the electrons that have already been waiting there. When you pull a plug out of the wall, the force field can no longer get in.

This, finally, answers the conundrum about why electrons don't pile up in the phone of someone you're speaking to. It's because your speaking has never been pouring them into the listener's phone in the first place! When you phone someone, all you're doing is sending along an invisible force field, which shakes the electrons that are *already* waiting in your listener's phone. The individual electrons barely travel—in fact, they drift along so slowly, barely at walking speed, that it would take over a month for a single electron to stumble all the way along a wire from New York to Los Angeles. But the weightless force field that makes the electrons jostle streaks that distance in a fraction of a second, making our seemingly instantaneous phone conversations possible.

Cyrus Field was always polite to Thomson after the success of the Atlantic cable, but seems to have steered clear of discussing invisible fields. It was still too bizarre for a businessman brought up in the era of clanking steam–engine technology to believe. Thomson, however, was convinced that electricity would be a big industry someday and would need convenient labels so that people could know how much of the pushing force that came from the invisible field they were purchasing. He probably would have wished to name that pushing force after his idol Faraday, but French officials dominated scientific naming throughout the nineteenth century, and although they had nothing against the esteemed Mr. Faraday, he did suffer the great misfortune of not being French, and also—it was almost embarrassing to have to bring up something so hurtful—he hadn't been fluent enough to publish his original findings in their language either.

Memoranda were exchanged, there were snide letters and backroom politicking, and finally, at a conference in Paris—with a most unhappy William Thomson in attendance—the official word for the strength usable from the invisible field was decreed. When Napoleon had invaded Italy decades before, many Italian patriots had been appalled, but Alessandro Volta—he of the two metal discs in his mouth, and the first steady battery— had understood that there was a difference between purity of convictions and worldly advancement. He'd embraced the French invasion, he'd commented with flowery elegance on the graciousness of Napoleon the wise liberator, and that— combined with judicious publications in French—meant he was the one the French favored now. Although Volta had never had a clue to why his battery worked, the intensity of the pushing force that flies into our homes is measured in "volts," not "faradays." When you buy a computer that says it's to be operated at

110 volts, that means it's designed to work when a force field rushes in that has been configured to provide 110 units of that pushing force.

At that point it might seem the story of electricity is over. There are ancient charged electrons hidden inside all matter, and there are force fields that can separate out those charged electrons and make them move. But if that were all, today we would still be living in a world of grand Victorian technology; there would be elaborate lightbulbs and telegraphs, and possibly even electrically powered horseless carriages, but that would be it. There would never have been radios or televisions or cell phones; no satellites beaming down GPS signals; no WiFi or Bluetooth or any other wireless technology. Yet even in Thomson's mid-Victorian times, there was a hint that a further aspect of electricity existed.

While the great Atlantic cable–laying was first under way, another friend of Thomson's, James Clerk Maxwell, had begun to look more closely into the fields that Thomson was trying to control. (He was the young scientist to whom Faraday had written, asking for help, in 1857.) He realized the fields had a complex inner structure, and were actually made of two parts—an electric part and a magnetic part.

His vision was extraordinary.

Every electrically charged particle in the universe forms the center of a huge force field, he wrote. This "electric" field stretches outward like a streaming aura. We all carry these vast auras with us as we move: they travel in perfect pace with us. Normally the positive and negative charges around us balance so that we don't see any effect. But if you scuff your feet on a rug and pick up an excess of negative electric charge from the electrons on the carpet, then the field emanating from you is denser, more intense. Lift your finger, and this slightly stronger field spreads outward, like the light from the Statue of Liberty's lamp. We can easily get static shocks—but there's also something more.

Whenever you shake that charged-up finger, the effect is like shaking a big bowl of Jell-O, or jostling your hand in a pond of water. The streaming field that stretches from you begins to wobble.

Now for the magical bit. If you take your jostling hand away from a real pond, then pretty soon the water's ripples die away. The wave you've created will stop. But Maxwell realized that there was also a magnetic part to Faraday's invisible force fields. As the electric part started rippling, it would power up this second, invisible magnetic part. (Why? Because changing electric fields produce magnetic fields—it's what Joseph Henry had realized with his electromagnets: switching on the electric current made the magnetic force appear.) But what then happened as the magnetic part of the invisible field became stronger? Well, changes in magnetic fields will make a fresh *electric* field pop into existence. That was what Faraday himself had shown in the great basement experiment in 1831.

What it all means is that just by jostling an electric charge, you will make an electric field begin to sway, and when those first ripples die down, they will make a fresh magnetic field appear. As that magnetic part dies down in turn, the change in *its* intensity makes a fresh electric field appear. When that weakens there's another magnetic field, and . . .

It never ends. You have to pour in some energy at the start to get the initial electric charge to shake, but once you've done that, once you've made the first of these mutually connected fields start to wobble, you can walk away. Decades and millennia might pass, our earthly existence might be totally forgotten, but the forward-stretching ripples you started in this combined "electric-plus-magnetic" force field will keep on traveling. It is immortal. The ripples are the magic carpets flying through the heavens; they "weave a web across the sky." Thomson had missed it, because he'd been confusing all this; he hadn't been able to see clearly the two separate parts and how they could create each other, phoenixlike, forever. But Maxwell had finally shown why Faraday's vision in his basement laboratory had been true.

Maxwell wrote up his ideas in a number of intricate equations. But when he died, in 1879, although Thomson was a supporter, Maxwell's ideas were still taken by most scientists as just a hypothesis. Who could believe we live in a universe swarming with such invisible waves? It was accepted that a few soaring waves might leak out here or there from ordinary electrical wires. But no one was actually trying to reconfigure those wires so that they became dedicated launch platforms for these ripples in the "electromagnetic" field; no one, even as the 1880s began, was building a detector to monitor where they might be landing.

And then, in 1887, someone did. The experimenter who accomplished this—Heinrich Hertz—left such a revealing diary that it's possible to present his work almost entirely through his own words, interwoven with the words of his contemporaries and immediate successors. It's to those firsthand accounts of the birth of our wireless world that we shall now turn.

PART III

WAVE MACHINES

6

A Solitary Man

Extract from the diary of Heinrich Hertz, 1884:

27 January 1884. Thought about electromagnetic rays.

11 May. Hard at Maxwellian electromagnetics in the evening.

13 May. Nothing but electromagnetics.

16 May. Worked on electromagnetics all day.

8 July. Electromagnetics, still without success.

17 July. Depressed; could not get on with anything.

24 July. Did not feel like working.

7 August. Saw from Ries's "Friction Electricity" that most of
what I have found so far is already known.

Letter from Heinrich Hertz to his parents, 6 December 1884:

There are rumors about my prospects here and at Karlsruhe, so
that I am asked by various [colleagues] "You want to leave us?"
while I know nothing at all.

Extract from the diary of Heinrich Hertz, 1885–1886:

28 November [1885]. Evening at the Saturday night social club.

13 December. Solitary morning walk through a snowstorm to Ettlingen.

31 December. Happy [that] this year is over, and hoping that it will not be followed by another like it.

22 January 1886. Bad cold and toothache.

23 January. Went to bed as early as possible after slinking about all day.

12 February. Worked all day on [a] battery.

18 February. Zealous work on the battery.

24 March. Took on an apprentice mechanic in Martin's place. His first performance: he smashes the glass plate of the large electric friction generator.

15 June. Whitsun holidays; depressed by danger of war.

31 July. Wedding day.

16 September. Undecided about which project to begin.

Heinrich Hertz, keynote address at the Imperial Palace, Berlin, August 1891:

Outside our consciousness there lies the cold and alien world of actual things. Between the two stretches the narrow borderland of the senses. No communication between the two worlds is possible excepting across the narrow strip. . . . For a proper understanding of ourselves and of the world it is of the highest importance that this borderland should be thoroughly explored.

***Memorial appreciation of Heinrich Hertz (1857–1894),
by Max von Laue, Nobel Laureate:***
Now began the classic period of Hertz's life. . . . While experimenting . . . Hertz found something unexpected.

Extract from the diary of Heinrich Hertz, 1887:
3 June. Not much to be done, because of damp weather.
7 June. Experimented a little; feel listless, with no desire to work.
15 July. Started to fill the large battery.
18 July. Experiments with sparks from the battery.
19 July. Desire to work vanishes completely.
7 September. Started to work in the laboratory on rapid
     oscillations.
8 September. Experimented . . . to the heart of the matter.

Letter from Heinrich Hertz to his parents, autumn 1887:
My colleagues, as far as they think about it at all, will believe that I am up to my ears in optical experiments, but meanwhile I have been doing some quite different things. . . .

Hertz's mentor, Hermann von Helmholtz, had long before asked him to test Maxwell's predictions. Hertz now realized he could achieve this by creating an apparatus with two parts. The first was a transmitter, where an electric spark jumped back and forth between shiny metal balls. He hoped that the moving field rippling out from the electric charges in those sparks would generate the invisible waves that Maxwell had predicted.

The second part of his apparatus was a square wire hanger. That was the receiver. If he was right, invisible waves would fly out from his transmitter, travel across the auditorium, and reach this metal hanger. To confirm that they had arrived, he had cut a tiny gap in the hanger. If an invisible wave arrived, it would have to cross that gap and would produce yet another spark.

There was no wire connecting the transmitter and the receiver. If he saw sparks in the gap of the receiver, then he'd know Maxwell's waves had flown across the room.

Heinrich Hertz's keynote address to the German Association for the Advancement of Natural Science, Heidelberg,
20 September 1889:
The sparks [which have to be detected in the receiver] are microscopically short, scarcely a hundredth of a millimeter long. They only last about a millionth of a second. It almost seems absurd and impossible that they should be visible; but in a perfectly dark room they are visible to an eye which has been well rested in the dark. Upon this thin thread hung the success of our undertaking.

Extract from the diary of Heinrich Hertz, 1887:
17 September. The experiments are yielding very beautiful and
   complementary results.

19 September. Set up experiments on the relative position of the circuits, and sketched them.

25 September. Sunday. Worked hard on second sketch at home.

2 October. A little daughter born at 2:45 in the morning, to be named Johanna Sophie Elisabeth.

5 October. Began work again in the morning.

Heinrich Hertz to Hermann von Helmholtz (Germany's leading physicist), 5 November 1887:

I should like to take this opportunity to let you know, most honored Herr Geheimrat, about some experiments that I have recently successfully completed. . . . I have some misgivings about taking up your time but this paper deals with a topic that you yourself once urged me to tackle some years ago.

To Hertz's parents, from Elisabeth Hertz (his wife), 9 November 1887:

Heins [her name for Heinrich] sent off his manuscript to Prof. Helmholtz on Saturday, and on Tuesday received in answer to it a postcard from him, containing only the following words, "Manuscript received. Bravo! Will hand it to be printed on Thursday." Naturally that gave us great pleasure; moreover, Heins had already started on new experiments Monday, and when he came home in the evening, he told me that he had set up the apparatus, tested it, and within a quarter of an hour had again succeeded in the most beautiful experiments. . . . He simply pulls these beautiful things out of his sleeve now! Of course . . . I certainly understand nothing of it.

***Memorial address by Professor Max Planck, delivered to the
Physical Society of Berlin, 16 February 1894:***
What research scientist does not remember even today the
feelings of wonder and amazement that came over him at the
news of these discoveries? Paper followed paper in quick succession, piling up new observations. We learned that electrical
processes can also produce (dynamic) effects; that electromagnetic waves travel through air; that electric waves propagate
entirely in the same way as light waves. And the proof for all
this was obtained through tiny little sparks, which must be
looked at in partial darkness with a magnifying glass in order
to observe them at all!

Heinrich Hertz to his parents, 13 November 1887:
This week I have again had good luck with my experiments. I
have never before been on such fertile soil, prospects are opening
right and left. The very thing that I have now done had been on
my mind for years, but I did not believe it could be realized . . .

Extract from the diary of Heinrich Hertz, 1887:
16 December. Went back to my experiments again and started
    to fill in the gaps.
17 December. Experimented successfully.
21 December. Experimented all day.
28 December. Experimented and observed the effect of the
    electrodynamic waves.
30 December. Followed the effect throughout the auditorium.

31 December. Weary from experimenting. Evening at the house
of my parents–in–law. Looked back on the year with
pleasure.

**Heinrich Hertz's keynote address to the German Association
for the Advancement of Natural Science; Heidelberg,
20 September 1889:**
All these experiments in themselves were very simple, but
[when I finished them] . . . it was natural to go a few steps
further. . . .

Heinrich Hertz to his parents, 17 March 1888:
I've [now] had the large chandeliers removed from the auditorium to give me the largest possible empty air space. . . . Yesterday I conducted some new experiments.

Memorial appreciation of Heinrich Hertz by Max von Laue:
The [next] investigation . . . was a great step forward. Whereas
transmitter and receiver had until then been kept in close
proximity, Hertz now separated them to the full extent permit-
ted by the largest room at his disposal, the auditorium of his
Institute, fifteen meters. The transmitter stood at one end wall;
the opposite end wall had been turned into a mirror for electric waves.

With the great distance between his transmitter and receiver, Hertz hoped to make the invisible waves bounce off the room's walls.

Extract from the diary of Heinrich Hertz, 1888:
27 February. Prepared for new experiments. Made metal
    shields.
5 March. Experimented with the formation of shadows by
    electromagnetic beams.
9 March. Death of the Kaiser.
14 March. Evening lecture in the Mathematics Club.

Heinrich Hertz, Collected Papers, 1894:
I thought that I noticed a peculiar reinforcement of the [waves] in front of . . . the walls of the room. . . . It occurred to me that this might arise from a kind of reflection of the electric force . . . the idea appeared to me to be almost inadmissible—so utterly was it at variance with the conceptions then current as to the nature of an electric force.

Hertz found that the waves he created, though invisible, could be reflected and bounced off mirrors, just as visible light could. They were all still created by the rippling patterns in the force fields stretching from the fast-moving sparks in his simple transmitter.

Extract from the diary of Heinrich Hertz, March 1888:
—Repeated the experiments with the utmost care.
—Experimented and thought I detected standing electro–
    magnetic waves by reflection in the hall.

Heinrich Hertz to Hermann Von Helmholtz, 19 March 1888:
I should like to report the following progress: electromagnetic
waves in air are reflected from solid conducting walls . . . the
phenomena are very pronounced and manifold . . . I have
also tried to project the effect to a greater distance by means
of concave mirrors and have obtained some indications of
success. . . .

To his parents from Elisabeth Hertz (his wife),
9 December 1888:
Today I write in Heinrich's place again; he is so engrossed in his
work that he does not want to interrupt it. . . . This morning
brought a letter from Geheimrat Althoff offering Heinrich a
choice between a professorship in Berlin and in Bonn.

Heinrich Hertz to his parents, 16 December 1888:
Bonn looks certain now; on the 22nd I shall have a conference
there with Althoff to settle the terms definitely. According to all
I hear, the lecture fees must make the holder of the chair a
wealthy man. . . . All in all, it cannot be denied that I have

proved to be in luck this time, at least outwardly and according to appearances. . . .

Hermann von Helmholtz to Heinrich Hertz, December 1888:
Honored friend, . . . Personally I regret you won't come to Berlin, but I must say I believe you are acting quite rightly . . . in preferring Bonn. . . . A person who yet hopes of grappling with many scientific problems had better stay away from the big cities. . . .

Heinrich Hertz to his parents, December 1888–January 1889:
I have . . . decided on the Clausius house. The garden contains what is already a beautiful chestnut tree. But there is a hitch.

Until four years ago these rooms served as a medical clinic, and although at the time the walls were scraped and new floors were put down, the medical people had nonetheless advised against moving into the apartment with a young family, since the apartment still could be contaminated. . . .

I went to see the professor of medicine (who) told me . . . there was little danger of contagion. . . .

From Helmholtz's nomination of Hertz to the Berlin Academy of Sciences, 31 January 1889:
The undersigned wish to introduce a motion to elect as Corresponding Member of the Academy Prof Heinrich Hertz. . . . Hertz [has] made himself well known by a series of very ingenious and unusually significant pieces of research. . . .

Extract from the diary of Heinrich Hertz, 1889:
17 March. Vigorously calculating.
26 March. Sent the paper off.

**Heinrich Hertz's keynote address to the German Association
for the Advancement of Natural Science, Heidelberg,
20 September 1889:**
[Electricity] has become a mighty kingdom. We perceive [it] in
a thousand places where we had no proof of its existence before.
. . . The domain of electricity extends over the whole of nature. . . .

**Heinrich Hertz to his parents, on an honorary visit to London,
arranged by the Royal Society, 5 December 1890:**
I left here on Friday evening and arrived in London at noon on
Saturday. . . . I was introduced to nearly everyone—and frequently did not get the names quite right . . . I was of course
especially fascinated to meet the older of the foreign physicists,
Sir W. Thomson and others. . . .

**From William Thomson's introduction to Hertz's
Collected Works:**
During the [many] years which have passed since Faraday first
offended physical mathematicians with his curved lines of force,
many workers and many thinkers have helped to build up the
nineteenth–century school. . . . Hertz's electrical papers, given to
the world in the last decade of the century, will be a permanent
monument.

Extract from the diary of Heinrich Hertz, 1891:

14 January. A little daughter [born], mother and child
doing well.

16 January. Charged the electrometer.

18 January. Tried a condensor for the transmission of electricity.

Heinrich Hertz to his parents, 1892:

Our days are now as peaceful as they could conceivably be. . . .
Unfortunately it has been spoiled as thoroughly as possible for
me personally, and thus for Elisabeth as well, because I have
caught a cold in the head, God knows how, which is as stubborn
as it is unpleasant.

Extract from the diary of Heinrich Hertz, 1892:

10 May. Installed a large sand pit in the garden for the children
to play in during Elisabeth's absence. It has a magic cave
in it.

27 July. My cold is becoming vicious. Broke off work and let
it lie.

**Memorial address by Professor Max Planck, delivered to the
Physical Society of Berlin, 16 February 1894:**

While at first his malady was thought to be harmless, there was
no notable improvement with treatment; rather, his difficulties increased as time went on. By the beginning of the winter,
[his friends] would not, could not, face the possible outcome. . . .

Extract from the diary of Heinrich Hertz, 1892:

29 August. Parents arrive on the way back to Hamburg.

6 October. Major operation.

7 October. Difficulties in swallowing very severe.

9 October. Severe pain.

11 October. Tried to get up, but fever very high.

Heinrich Hertz to his parents:

Now unfortunately my strength has been numbed for some time to come. . . . But I still hope for a future time when I shall be able to concentrate completely.

Heinrich Hertz to his parents, October 1892:

Unfortunately I cannot report anything encouraging about my-self; there is absolutely no progress and the only consolation—if it is a consolation—is that according to experience these condi-tions tend to be protracted. . . .

Extract from the diary of Heinrich Hertz, 1892:

19 October. A troubling period.

28 October. The swelling behind my ear keeps growing; unsuccessful attempts to drain through there.

29 October. Walb called Professor Witzel, who performed operation chiseling through mastoid.

Heinrich Hertz to his parents, 23 December 1892:

Is Christmas really here? It seems only yesterday that it was the middle of summer, and since then I have not been aware of anything that has happened, or that I have experienced anything but a dreadful dream from which I still cannot awaken.

Miscellaneous diary entries of Heinrich Hertz, Bonn, 1890s:
—Took a walk, searched in vain for starting points for fresh work.
—Experiments with the iron barometer pipe.
—The experiments yield probability of favorable results.
—The experiments seem to show little promise; therefore
    discouraged and broke off same.
—Felt fed up with work in physics.

Heinrich Hertz to his parents, December 1893:
I am among those . . . destined to live for only a short while. . . .
I did not choose this fate, but since it has overtaken me, I must
be content.

Heinrich Hertz succumbed to what was then termed blood poisoning on January 1, 1894, possibly caused by infection from substances in the medical clinic that had previously occupied his house. He was thirty-six.

Soon after, his theoretical researches began to be taken up by practical inventors, most notably the son of the Irish-born heiress to the Jameson whisky fortune. Since she had settled in Italy, her son—though fluent in English—was known by her husband's name: Marconi.

From Guglielmo Marconi's lecture accepting the Nobel Prize, 11 December 1909:

At my home near Bologna, in Italy, I commenced early in 1895 to carry out tests and experiments with the object of determining whether it would be possible by means of Hertzian waves to transmit to a distance without the aid of connecting wires. My first tests were carried out with an ordinary Hertz oscillator [which used sparks to generate waves in a manner similar to Hertz's early experiments]. With such apparatus I was able to telegraph up to a distance of about half a mile.

In August 1895 I discovered a new arrangement. . . .

From Submarine Telegraphs: Their History, Construction, and Working, *by Charles Bright, 1898*:

Marconi has succeeded in making the resonator work . . . at a distance of nearly nine miles. It is . . . the very latest thing in Inductive Telegraphy.

Marconi's wireless telegraph is based on the principle of turning Hertzian waves to account in transmitting them . . . by means of electric sparks. . . .

From "Signalling Through Space Without Wires," *a public lecture by Sir W. H. Preece, the Royal Institution, London, 4 June 1897*:

In July last year Mr. Marconi brought to England a new plan. Mr. Marconi utilizes electric or Hertzian waves . . . he has invented a new relay. . . . Excellent signals have been transmitted across the Bristol Channel.

Proceedings of the Royal Society, 28 May 1903:
The remarkable success of Marconi in signalling across the At-
lantic suggests a bending of the waves round the protuberant
Earth.

The signals that Marconi sent were simply more powerful
versions of the invisible waves—the undulations in the electric
and magnetic fields—that Hertz had produced in his laboratory.
Since they "radiated" outward, they stopped being called Hertzian waves and came to be called "radio" waves.

From the Encyclopaedia Britannica, *Eleventh Edition, 1910*:
High power stations are now used for communicating across
the Atlantic, and messages can be sent by day as well as by
night. . . . Hertzian wave telegraphy, or "radio-telegraphy," as
it is sometimes called, has a position of the greatest importance in connexion with naval strategy and communication
between ships.

**Message received from SS Olympic, 1,400 miles off
U.S. East Coast, 14 April 1912:**

... ... / — .. — — .— —. .. —.—. / .—. .— —. /
.. —. — ———— / .. —.—. . —.... . .—. ——. /
... .. —. —.— .. —. ——. / ..—. .— ... —

(The message reads "SS Titanic ran into iceberg. Sinking fast." It was received by a young telegraph operator, David Sarnoff, at Wanamaker's department store in Philadelphia.)

Many inventors and other interested parties began to imagine fresh ways of using this new device.

Memorandum from David Sarnoff, former telegraph operator, to Vice-President Edward J. Nally of the Marconi Wireless Telegraphy Company of America, 1915:
I have in mind a plan of development which would make radio a "household utility" in the same sense as the piano . . . the idea is to bring music into the home by wireless. . . . The receiver can be designed in the form of a simple "Radio Music Box" and arranged for several different wave lengths, which should be changeable with the throwing of a single switch or pressing of a single button.

If only one million families thought well of the idea, it would . . . yield considerable revenue.

Sarnoff's memo was rejected. A decade later, in the 1920s, sales of radio apparatuses by the company he founded—RCA—had made it one of the most powerful industrial firms in the world.

Radio transformed every country where it was introduced. Although telegraphs and telephones sent messages at extremely high speed, they still could only link one individual with another. Radio waves, however, weren't locked within a narrow copper wire. Since the physical nature of those waves is to spread in all directions, radio sent its information so widely that the term *broadcasting* became the popular way to describe this new effect.

Suddenly, national brands became far more popular to shoppers than they had before; local sports teams increasingly attracted a nationwide fan base; the cult of celebrity—as with Hollywood stars—became even more widespread. Listeners felt that radio broadcasts were being aimed personally at them.

And politics changed as well.

Adolf Hitler, Mein Kampf:
All propaganda has to appeal to the people and its intellectual level has to be set in accordance with the receptive capacities of the most-limited persons among those to whom it intends to address itself. The larger the mass of men to be reached, the lower its purely intellectual level will have to be set. . . . Even from the most impudent lie something will always stick.

In America, the demagogues who used radio to spread their message always remained in a minority. But in Japan and several European countries it was different: the Nazi Party's innovative use of radio transmissions was a major factor in its electoral success in the years leading to 1933.

As the 1930s went on, German tank commanders began practicing with radio command of large armored and aerial formations they could use to destroy neighboring countries. A few researchers in other countries began to wonder if the power embodied within radio waves could be enough to keep enemy war machines at bay. The British government especially wanted to know. The Royal Navy had long protected the country by sea. Would it be possible to use radio waves to protect it by air?

From A. P. Rowe, Secretary of the Committee for the Scientific Survey of Air Defence; to H. T. Tizard, FRS, Rector of the Imperial College, 4 February 1935:

Dear Mr. Tizard,

A copy of a secret memorandum prepared by Mr. Watson Watt on the possible uses of electro magnetic radiation for air defense, is enclosed herewith. . . .

7

Power in the Air

Many technicians had stumbled on radar over the years, but the first ones had found it impossible to get their superiors to believe them. In September 1922, for example, Albert Taylor and Leo Young of the U.S. Navy were trying to send a simple radio signal across the Potomac River, but kept getting some sort of interference. They looked up; a steamship was in the way. Yet when they tried to get funding to investigate this effect, they were scoffed at: how could a bulky steamship have any effect on ghostly, weightless radio waves? Similar effects were reported in Russia, France, and most other places where radio was used a lot, but the response was almost always the same.

It didn't help that radio technicians tended to be quiet sorts. But, luckily for the survival of Britain, and indeed of the civilized world, at least one radio expert was overwhelmingly the opposite in personality. His name was Robert Watson Watt, and in 1935 he could be found working in the dreary reaches of the atmospheric research station of the National Physical Laboratory, near the equally dreary English town of Slough. The poet John Betjeman loved the south of England, but after getting to know Slough, even he was inspired, famously, to write:

> *Come, friendly bombs, and fall on Slough*
> *. . . and blow to smithereens*
> *Those air-conditioned, bright canteens . . .*

A direct descendant of the James Watt who'd perfected the steam engine, Watson Watt had been a promising university student in Scotland, but things had never quite worked out since then. His marriage had long since tapered off to boredom ("I was . . . a dull, drab partner in the small residuum of the twenty-four hours that were not monopolized by work and sleep"), and pretty much everything else was on the same level ("I am five foot six, tubby if you want to be unkind, chubby if you want to be a little kind; a bit of a meteorologist . . . thirty years a Civil Servant.")

That last item was the problem. Given his great family legacy, Watson Watt had never imagined he'd end up middle-aged, middle-income, and not even of middling fame, there on the edges of governmental research departments, miles from London.

And then, in January 1935, a request fell his way, from heaven, or rather—it seemed as good—from the Air Ministry in London. A contact there asked him if there was any truth in the rumor that evil "death-rays" could be sprayed from radio transmitters at an aircraft. The question itself was easily enough answered in the negative, for radio waves are too weak to damage a hulking aircraft. But Watson Watt wasn't about to let the matter drop so easily.

He knew that the moment he sent his reply back, the door that was now briefly open to the London Ministry would close; he'd be stuck in Slough, perhaps forever. But if he played with the idea and came up with something better, then—who knew?—he might be regularly called to London. There would be all-expenses–paid day trips on the railway; briefings he could modestly deliver; superior people to meet; possibly a promotion.

What actually resulted from his response to the casual inquiry—ultra–top–secret missions to Washington, private briefings for Churchill; a knighthood from the Queen, and vast funds from a victorious nation—was beyond any of his imaginings. But for the moment, in this wet January of 1935, he actually had to come up with something interesting to tell the Air Ministry in London, and this presented a problem. Although Watson Watt was very proud, he was also rather honest with himself. He understood that although he was a good enough meteorologist, he was, in his sad estimation, at best "a second–rate physicist" and "a sixth–rate mathematician."

He'd made friends, however, with a colleague in his office, Arnold Wilkins, who hadn't been at Slough long enough to resign himself to second–rateness. Wilkins was keen to calculate what else might happen when a radio wave was sent out in the direction of an incoming aircraft. The invisible waves that Faraday and Maxwell and Hertz believed in—those magic-carpet-like undulations in the force field—wouldn't carry enough power to melt the plane or injure the pilot. But could they do something else?

Wilkins thought about it and realized they provided a way to use an enemy's airplane *against* itself. From his original training in physics, Wilkins knew something important about metal and, in particular, about what could happen inside the metal of an airplane's body. Watson Watt had some training in this area, but wasn't as deft with calculations, which is why Wilkins led the way.

Back in the time of Faraday and William Thomson, only a few theorists had imagined that invisible waves could affect ordinary solid substances and make them move. This was fair enough, for when there are no separate charged particles available—when all the charges are tucked away in balanced groups inside an atom as in the ordinary objects around us—then the electric and magnetic forces have nothing to latch on to. (Gravity, by contrast, doesn't have opposing parts that can be balanced into neutrality, and so is always noticeable.)

By the time Wilkins and Watson Watt were collaborating, though, it was becoming clear how the electric effects might work. In the 1930s an atom was popularly thought of as a miniature solar system, at the center of which is a big, heavy nucleus, like our sun. On the outside, spinning in distant orbits, the separate electrons are like our planets. Radio waves are just undulations in a stretching electric and magnetic field, so when a radio wave or the like swirls over one particular atom, it tries to tug loose some of those electrons.

Often a wave has no effect. Since the electrons in our bodies are generally held pretty tightly to the nuclei at the center of our atoms, our bodies are invisible to most of those fields. A radio wave will hurtle right through us; to radio waves, we are ghosts. Even the atoms inside ordinary rock or bricks are constructed in such a way that radio waves fly through, which is why we can use a cell phone inside a house.

Metal is different. The atoms in iron or aluminum are more loosely constructed, and are like solar systems that don't particularly care about their outermost planets. Although the majority of these atoms' electrons stay in orbit, the outermost ones are free. A strip of shiny aluminum with several billion atoms can be viewed as a galaxy of several billion stars which all have some planets orbiting close, but have let the outer ones escape. It's as if innumerable electrically charged Neptunes or Plutos were floating loose among the stars in that galaxy, joined by similar refugees from other solar systems.

That's what it's like in the metal wing of an airplane. When a radio wave storms into one of these miniature galaxies in an aircraft's metal, the innermost electrons of each miniature solar system might be buffeted a little, but they won't be knocked very far. But the more distant electrons, the errant, solitary, orphaned electrons, the ones flying loose in the miniature galaxy, are a different story. The radio wave that flies through the metal "nabs" them, and carries enough power to start tugging them along.

When this happens inside a tiny metal receiver like the one within a cell phone, the solitary electrons start wobbling, that wobble is magnified, and our crucial information—such as "Hey, I'm in the car!"—is transmitted. But Wilkins realized that when a radio wave hits a much larger expanse of metal, the effect is more dramatic.

In an enemy airplane there are yards and yards of such vulnerable, waiting metal. Any radio wave we send in its direction accelerates the loose, mobile electrons there. And each electron is always surrounded by its own personal force field. If the electron is left still, that force field will be fairly still and no signal will be produced from the airplane's wing. But when the electron is made to ripple from side to side, that force field ripples as well. (This was what Maxwell and Hertz had realized.)

By aiming a radio transmitter at the targeted plane, trillions upon trillions of electrons could be made to ripple in unison and serve as tiny bubbling radio transmitters of their own. In other words, by pumping up invisible radio waves, Wilkins could force the enemy plane to become a flying transmission station! The entire airplane turned into an antenna that could not be switched off.

The big question, though, was whether the transmission would be powerful enough to be detected. For the sky is big, but radio waves are small. Most of the radio waves beamed outward would disperse and miss the airplane, or be very weak by the time they reached it. Wilkins did the calculations. Even though the launched wave might disperse so widely that only one thousandth of its strength would be available to tug on the metal electrons in a plane four miles away, that would still be enough. Once again, it was the very minuteness of electrons that allowed this to work. Wilkins showed that even that diffuse a wave would make about 60 quadrillion (60,000,000,000,000,000) electrons start rushing back and forth each second at any given point in the airplane's wing. The invisible radio waves generated by those rushing wing electrons would be powerful enough to detect back on the ground. (Stealth aircraft, invented decades later, avoid detection partly because the paints used don't allow much incoming radar energy into the plane's metal skin, partly because the aircraft's surfaces are angled so that anything it does reflect will aim away from the original transmitter.)

Watson Watt had supervised Wilkins's work, sort of, so it was fair enough that when the memo went back to the Air Ministry in London, Wilkins was generously mentioned as an important collaborator. But it was Watson Watt's name, alone, that went on top.

Wilkins didn't mind, he was happy to take a backseat role, for he knew something about Watson Watt that the mandarins in Whitehall were about to experience. For Watson Watt loved to talk—no, not just talk, he loved to expostulate, to verbally cogitate, to burble and engulf and generally overpower anyone he was near with a torrent of words. At a court case about patent rights, where his substantially ad-libbed remarks totaled a third of a million words, Watson Watt remarked that "it would be disingenuous to leave any impression that I did not enjoy this experience."

Now, eager to push this idea that might free him from Slough, he went into overdrive. He sent reminders to London, and then a second memo ("I have thought it desirable to send [this] immediately rather than to wait . . ."). Watson Watt met the mild, pipe-smoking civil servant A. P. Rowe, who'd initiated the query; then lofted higher, for lunch with Rowe's superior, Henry Wimperis, at the Athenaeum Club. Pretty soon, senior individuals—*very* senior individuals—were interested in what Watson Watt was proposing.

It helped that Britain had few alternative methods for defending against German air attacks. In World War I, blind people with excellent hearing had been placed under the likely path of incoming Gotha bombers and asked to don stethoscopes attached to large Victrola–like horns. More recently, in the early 1930s, a giant concrete "ear," two hundred feet long and twenty-five feet high, had been built on marshes near the Thames, aiming over the Channel, where possible enemy aircraft might emerge. Neither of those techniques worked very well. The only chance of succeeding was to create wobbles in the invisible electric force fields that stretched out from the charged electrons in transmitters. Those waves would continue wobbling outward, possessing enough power to kidnap and redirect the electrons in the metal of incoming planes.

It sounded like science fiction, but within three weeks, Watson Watt had convinced Wimperis, Rowe, and everyone else within earshot that the time was ripe for a full-scale test. He wouldn't need any special equipment. The "enemy" plane could just be an ordinary RAF bomber; the transmitting radio station could simply be one of the powerful BBC Empire masts at Daventry in Northamptonshire, which was already broadcasting regularly; the oscilloscope to detect the waves that would surely be flung back out by the bomber's electrons could be borrowed from a research colleague.

At this point Watson Watt knew he had not just to convince the scientific staff at the ministry—who would be tolerant of teething problems—but also the operational leaders, and in particular, the extremely suspicious Air Chief Marshal, Hugh Dowding, known even to his friends as "Stuffy." A great number of weapons that promised seemingly magical power for the defense of Britain had come Dowding's doubting way, and all of them had failed operational tests. His resources were so slight that he couldn't gamble on backing the wrong one.

Early on February 26, 1935, the RAF bomber flew past the assembled watchers on a grassy field near the BBC transmitter. (Watson Watt had quietly added a very long dangling strip of metal to the plane to be sure the demonstration worked.) The bomber was tracked to a distance of eight miles. When the test was over, Watson Watt turned to a ministry official, and—in what seems his one recorded moment of brevity—said, "Britain has become an island once more."

Dowding didn't know that the first test was rigged—he was as familiar with the theory of electrons within metals as he was with medieval vowel forms in Serbo-Croat—but he did know that this pudgy, absurdly confident man from Slough was making things happen that no one had made happen before. The RAF had a tight budget, and fighter command had one of the smallest portions, but even so, within a few weeks Dowding had shifted the equivalent of more than one million dollars to Watson Watt. His sole assignment was to find out more about how this "radar" could work to Britain's advantage, and then show how to build practical radar stations. (The word *radar* was not in fact coined until 1941, when two American naval officers came up with it as a convenient shorthand for "radio detection and ranging." In 1935 the less revealing label "Radio Direction Finding" was used, so as not to give too much information to the enemy.)

But despite the RAF's support, Watt, like any outsider, needed a stronger protector in the government bureaucracy. As his first memos floated along the corridors of Whitehall, they ended up, luckily, being noticed by the kindly Henry Tizard, a superb administrator. Tizard had test-flown World War I Sopwith Camels, been a lecturer in thermodynamics at Oxford, then the head of Imperial College, and—most usefully for the bureaucratic battles to come—an excellent, scrappy lightweight boxer in his youth.

There was an especially precarious moment for the project, when Parliamentary twists briefly gave Churchill more influence on the government. Churchill was all for increased defenses against Germany of course, but when it came to science, he depended completely upon the advice of Frederic Lindemann— an angry, status-obsessed ex-academic who could be serenely charming when he wished, and who had the knack of making upper-class individuals feel that they were as wise as the greatest thinkers. Since Churchill's own scientific education hadn't quite reached the levels of the early nineteenth century, he had no way of recognizing Lindemann's incompetence.

Churchill pushed Lindemann onto Tizard's committee, and Lindemann immediately explained that he knew for a fact the newfangled radar defenses they were planning were never going to work effectively. It didn't help that when Lindemann and Tizard had been young researchers, visiting in Berlin in 1908, they'd once agreed to settle a point of honor in the boxing ring. Lindemann was a much larger man and couldn't bear it that the wiry Tizard had pummeled him; he refused to shake hands afterward. Now in London, Lindemann slowed the constructions that Watson Watt was preparing for several months until, through still more deft bureaucratic footwork, Tizard managed to get Lindemann expelled. He created a dangerous lifelong enemy, but for the time being he had also cleared the way for Watson Watt to proceed.

By the time of the Munich Crisis of 1938, as the screaming voice of the Führer burst out of German radios, and the signals carrying his words rippled outwards—just as Michael Faraday might have predicted a century earlier—some of those radio signals were soon floating over five huge constructions in the south of England. These were the first of the Chain Home radars: most of the stations had tall metal transmitters, like massive electricity pylons, 350 feet high; the receiver pylons were 250 feet high, and generally built of wood. By the summer of 1939 there were twenty similar stations, concentrated in the southeast of England, but with a few scattered all the way up into Scotland. From each transmitter, long, undulating waves sped out. During the Munich Crisis, the transmitters had sent those waves to follow Prime Minister Chamberlain's plane to a distance of nearly a hundred miles from the British coast, long after it was out of visual range. Now, with war approaching, the transmitters were increasingly used to identify equally distant British planes, simulating the paths that German fighters and bombers might take.

The German armed forces had some hints of what was happening, but never got a complete picture, for the various divisions of the German government didn't talk to one another much. When a Luftwaffe officer named Colonel Wolfgang Martini did finally decide to probe Britain's defenses in the last peacetime months of 1939 (by trying to listen to any radio signals a radar network would be producing), he most usefully for Britain resolved to be extremely thorough. This meant he had to carry a lot of heavy equipment, and so he used a giant zeppelin airship for the job. (It was the *Graf Zeppelin* LZ–130, sister ship of the LZ–120 *Hindenburg*, which had exploded in America just two years before.)

He couldn't have chosen a better way of beaming his location to the British. Zeppelins were coated with shiny aluminum varnish, and aluminum is a metal, which meant that the surface of the zeppelin consisted of hundreds of square feet of loosely attached electrons. It formed a perfect target, suspended conveniently high in the air. The first signals from Britain's radar stations whapped into the zeppelin even before it had completed crossing the English Channel. Electrons in the zeppelin's aluminum–painted skin immediately began shimmying from side to side, and so began pumping out miniature radio signals just as Wilkins had predicted. While the zeppelin lumberingly hid in clouds during its reconnaissance mission up and down the coast, most of the high masts of the Chain Home stations stayed quiet, giving away no trace of their operation. When the final Luftwaffe intelligence report on British preparedness was delivered, in July 1940, it said nothing about the Chain Home system.

By the summer of 1940, Germany's army had conquered most of western Europe and was perched on the edge of the English Channel, just an achingly few miles short of Britain itself. The German navy was assembling a huge invasion fleet, and if that got across the Channel successfully the few armored British units that survived would be heavily outnumbered. (Many British tanks had been destroyed or left behind in France.) Churchill would likely be executed, if he didn't escape to Canada; German administrators had lengthy lists of prominent Jews, trade unionists, clergy, and other undesirables who would be killed too.

The Royal Navy would do what it could to block the invasion, but without command of the air, there was little chance it could succeed indefinitely. The RAF was the crucial barrier to Germany's advance—but it seemed feeble indeed. Spending had been low for Fighter Command, so although there were a good number of topnotch planes available, there weren't enough trained pilots to keep defensive patrols circling in the air. Hitler had only to get rid of this limited RAF and then he could invade. The air attacks weren't expected to take more than a month, top German officers calculated. They would begin on August 13, 1940—*Adler Tag*, or Eagle Day.

Around 5:30 a.m., the radar operators in the easternmost radar stations in Britain detected a German air formation assembling over Amiens in France. Soon there was another one over Dieppe, then a third north of Cherbourg. The German pilots had every reason to think they were unobserved, for they had no way of detecting the electric and magnetic waves that surged through the metal wings of their planes and produced the signals that—only a fraction of a second later—were accurately received back in England. But when those several hundred planes did reach English airspace for their surprise attack, ten RAF squadrons were waiting for them. As a contemporary observer put it, "Inhospitably, the reception committee turned all these overseas visitors away."

It was Watson Watt's finest hour. Time after time, in the weeks after that, Germany tried to launch a surprise attack, so it could take advantage of its superiority in numbers of planes. Later in August, for example, the German air force feinted with a huge attack on the south of England, while sending an even greater, secretly assembled air armada over from Denmark to attack what it was sure now had to be the undefended northeast. But when those attackers were still far over the North Sea, the RAF—with deadly pilots from Poland, Norway, and throughout the Commonwealth—swooped upon them.

It helped that when a few suspicious German intelligence officers did promote attacks on what they thought might be radar stations, the British operators, mostly women, stayed calmly at their oscilloscopes, despite substantial casualties. There was no change in the RAF's responses. The German High Command was confused enough to conclude that perhaps the curious high masts weren't a central part of the British defense after all.

It also helped that the British government encouraged alternate cover stories about the defense scheme. There were leaked press reports that the accurate intercepts, especially at night, were due to the vision-enhancing properties of carrots (which seems to be where the story began that carrots are especially good for your eyes). There was also quick thinking by RAF members. Sergeant Phillip Wareing was shot down while chasing an enemy plane back to France and interrogated shortly after his capture:

"One German asked me, 'How is it you're always there when we come?' I said, 'We have powerful binoculars, and watch all the time.' They didn't query that at all."

Through the rest of August, and well into September, the radar stations continued directing one fresh group of RAF pilots after another straight to the fleets of attacking German planes. Streams of bouncing electromagnetic waves filled the sky, guiding the British planes forward, directed by readings from the calm survivors at the radar stations. Tons of metal wreckage fell from the air. The RAF suffered large losses, but the Luftwaffe's were larger still. By the end of September, autumn storms in the Channel had begun, and on October 11, Operation Sea Lion—the planned German sea invasion of Britain—was indefinitely postponed.

Yet the British lead in radar didn't last. As German night attacks on London and other large cities continued, British radar operators noticed something disturbing about the Luftwaffe's tactics. The German planes were taking up accurate positions more quickly than they had before, even in total darkness.

There was at least one occasion when German dive-bombers based in France flew directly to a British destroyer sixty miles from shore and sank it, even though it had been well out of visual range.

This was bad news. The Germans were not just catching up with the British in their use of radar, but surpassing them. Britain's Chain Home radars sent out waves many yards in length. That was good enough for getting a rough fix on airplanes coming in over the English Channel, but such long waves quickly spread wide, wasting much of their energy by uselessly bathing the cows, milk vans, and open fields of southern England. Watson Watt knew that a shorter wave would be easier to aim and more tightly focused. This must be, he and others realized, what Germany was now able to create. The fact that German officers could synchronize their plane movements at night and hit distant warships with great accuracy proved it. But where was Germany hiding the machines that poured out these small electric waves to such deadly effect?

Radar so far had been the hero—a protector of the nation. Yet once Britain learned how Germany was catching up, it would become a villain.

8

Power Unleashed

HAMBURG, 1943

The clue that allowed the new German radar sets to be found—and set in motion the events that led to Faraday's invisible waves being used for immense destruction—seemed innocent enough. It appeared in an eight-page typewritten document that had been sent by an anonymous German citizen to the British naval attaché in Oslo in 1939. The report seemed too outlandish to be true, for it described activities beyond anything Britain had contemplated. There were accounts of a science-fiction-like research establishment on a distant Baltic island where jet-propelled glider-shaped planes were built, along with descriptions of not just one but two radar systems far more advanced than anything Britain had. When the packet was translated and distributed in London, only one individual thought it worth keeping.

That individual, however, was Reginald V. Jones, a very young-looking twenty-eight-year-old. Although his official training was in astronomy and physics, he had done graduate work at Balliol College, Oxford, where a broad literary background was expected, and it was taught that every received opinion deserved to be questioned. He recognized that Nazi administrators were so suspicious of one another that it was quite possible—as indeed turned out to be true—that several German research teams had been working on radar without the main Luftwaffe officials being aware of it. Now, in 1941, he realized that Germany did have a functional radar system, and—from reading accounts of prisoner conversations and radio intercepts—he also knew that it was code-named "Freya."

To someone of his background, this code name carried a clue that would be as revealing of German intentions as if a carrier pigeon with the true information had been sent direct to the Office of Air Intelligence, Whitehall, London SW1. The German High Command was obsessed with Aryan myths. The solution to this great electronic danger of the mid-twentieth century, young Jones realized, would come by going back in time nearly one thousand years.

Jones strolled from his Whitehall offices, accordingly, one day late in 1941, to Charing Cross Road, the center of London's book trade. Before the day was out, he had found what he wanted. Freya was an ancient Norse goddess who was described as being usually in the company of Heimdall, another mythological being. Freya had a necklace, and Heimdall's job was to guard it. To do this, he was given the power to see a hundred miles in all directions—by day or night.

RAF reconnaissance flights had already found some installations in occupied France that looked as if they could be radar sites, but the devices had seemed too large and old-fashioned to be the cause of the most accurate interceptions now. No one, however, had thought there might be two machines, with different powers, working together. The reference to old Norse mythology—where Heimdall and Freya worked together—made Jones think that this might be what the German command was now building.

Jones sent a camera-equipped Spitfire fighter back to the first big site that had been found, a few dozen miles from Le Havre, near the town of Bruneval. It was highly polished for top speed, and it roared in just a few hundred feet from the ground, to keep down the amount of time that guards or antiaircraft guns would have to try to fire at it. As soon as the negatives were back in London, Air Ministry photo interpreters got to work. At first they saw only the troop emplacements and barbed-wire barriers and commandeered chateau that would be expected at an ordinary base, but a closer look with magnifying glasses revealed that where a path away from the chateau stopped short, there was indeed yet another radar device, tucked into a clearing no more than a few yards wide.

This was worse than expected. A radar set small enough to fit in this clearing couldn't have the huge antennae that Britain's Chain Home radars required. Instead the photos suggested that the German engineers had produced a radar with waves that wiggled out at five feet or perhaps even less in length. The entire device could fit on the back of a van, with a single yard-wide steerable antenna on top to generate and receive its waves. (Since the Chain Home antennas were the size of very tall electricity pylons, it would have taken hundreds of cubic yards to rotate them, and no engine known could easily do that.)

No one in Britain knew how Germany had achieved this feat of engineering. If there were more of these sets, and if they were crammed aboard a plane, then German patrols could identify vulnerable American troop transports coming to Britain even in darkness in the middle of the ocean, and send that information to waiting U-boats; individual Luftwaffe fighters could pick out Allied planes in total darkness as well.

Two volunteers from the French Resistance, Roger Dumont and Charles Chauveau, went in to investigate the base and its small, deadly radar. They found more than a hundred German troops, and at least fifteen machine-gun emplacements. The Royal Navy couldn't get up there to examine or capture the radar—code-named Würzburg—for although the chateau at Bruneval was on the coast, the sea had cut cliffs nearly four hundred feet high in the soft chalk. It had to be attacked, and since there were no landing strips, one of the newly formed paratrooper units would have to go in.

And that was how Charles W. Cox, a diminutive peacetime film projectionist, amateur radio ham, and now most junior sergeant in the Royal Air Force, came to be instructed, one February day early in 1942, to report to the Air Ministry in London. Air Vice-Marshal Victor Tait was waiting for him. Normally Cox tried to spend the day in carpet slippers—he had corns on his feet—and he tended to keep up his trousers with insulating wire rather than a belt. But today, he realized, was special, and he tried to dress properly.

"You've volunteered for a dangerous job, Sergeant Cox," Tait said.

"No, sir," Cox protested.

"What d'ye mean, 'No, sir'?" Tait asked.

"I never volunteered for anything, sir!" Cox said.

From an unpublished twenty-page account left by Cox's estate, it's possible to reconstruct what happened next. Cox was sent to the tightly guarded Ringway training facility near Manchester, where at first he couldn't understand why he was marching amid dozens of tough paratroopers. After all, he'd never been in an airplane and was afraid of exposed heights. Then in an awful moment he realized that he was in training with them because he was going to be in on their assault. Someone had to supervise the disassembly of the Würzburg radar, if it was going to be captured and brought to England, and on this short notice, his experience as an amateur radio operator meant he was the best the War Office could get.

Part of the First Airborne Division, the paratroopers were wearing army uniforms, yet Cox was in RAF uniform. If the assault team was captured, he would stand out, and the Gestapo would want to know why he was with them. Cox most understandably worried about this, and before long, Jones went up to Ringway to comfort him. When Cox explained that he didn't so much want comfort as the chance to blend in by wearing the same uniforms as the other troops, Jones had to reply that he'd tried in London to get that approved, but the War Office was adamant that switching service uniforms set a bad precedent. Cox explained that he understood about precedents, but that with the Gestapo on hand this could, with justice, be considered a special case. Jones, wincing, had to explain that the War Office had been *very* adamant.

After the rush course in parachuting, Cox—still in his RAF uniform—was bundled off to Salisbury Plain for an even more rushed course in assault techniques. Some of it he enjoyed, not least the way that anything he asked for was given to him— binoculars, a compass, clean boots, even a smart new Colt .45. Some of the course was more distressing, as when he found out how the Scottish troops he was training with practiced cross- ing a barbed-wire emplacement. Instead of bringing out wire- cutters, as he expected, they simply had one of the men stretch out on the uncut wire to form a human ramp: the rest of the men then crunched over him. None of the officers explained to Cox how difficult the operation was likely to be: they would have only relatively light weapons, and prior airborne opera- tions had proven far more difficult than expected.

The parachute assault took place on the night of February 27, 1942. The Würzburg set they were coming to get began tracking them when they were still twenty miles out, sending fast- rippling invisible waves up into the dark air. Answering electric waves from the incoming British planes shot back ineluctably, launched into the night from the metal wings and body of each aircraft, emanating just inches from where Cox and the para-troopers sat. With those unseen answers, the British team was detected.

When the approximately one hundred men jumped, about two dozen of the team were blown hopelessly off course, coming down miles away, but the bulk of the force landed safely in the target zone. After the usual ritual upon landing—blissfully relieving themselves of the tea they'd drunk in the hours before boarding their planes—the assembled troops walked quickly toward the chateau that overlooked the Würzburg radar. Cox followed, pushing the little wheeled trolley the planners in London had decided he should use to put the captured radar parts on.

Only a few of the defenders were at the chateau itself, but the German radar operator had understood why so many glowing dots on his screen were heading his way. The main group of German troops had been warned and were either in ambush positions or trying to get there. A series of firefights began, and at that moment Cox realized he had to move quickly. There were likely to be hidden explosive charges that could detonate the Würzburg, and Tait would definitely think badly of Cox if that happened, so now Cox hurried forward in the dark, amid the firing, and got over the barbed wire around the set to help find and disarm those charges. He recognized a suspicious moving shadow and thereby helped—most politely—to capture the German radar operator who was running away; he directed the British troops in the use of crowbars to snap loose electronic parts from the Würzburg, and calmly made sure they also collected the serial numbers of any replacement pieces. He was so worried by what might happen if he arrived back without enough samples that even as the German troops approached and mortar attacks began—and while the main paratrooper force had to rush from the Würzburg earlier than planned—he kept at work until he was sure key parts of the swinging curved antenna had been sawed loose.

There was supposed to be an easy scramble down to the beach, but the men who had been allocated to cover this escape were precisely the ones who had been blown off course on landing. A German machine-gun crew started firing at the British, which further worried Cox; as a result he directed the men to shift the Würzburg parts from the trolleys to backpacks (thus ensuring that none of the electronics were damaged), and he also seems to have helped tend the wounded. Just when it looked as if the entire assault team might be overrun, there were sudden shouts of *Cabar Feidh!* as the lost Scot Highlander covering party finally arrived. The German machine-gunners decided that escape into a convenient ravine was wiser than facing any more of these shouting troops. The raiding force at last made it down to the beach.

The Royal Navy wasn't there, however, and increasingly urgent radio messages and even bright flares didn't seem to help. But as headlights from German trucks appeared near the clifftop above, the navy's landing craft arrived, lots of them, and their heavier guns meant the fresh enemy troops above would be no problem. Most of the escaping troops were left to wallow all the way back to England in the unwieldy landing craft (though they were given rum once aboard), but Cox was taken aside and quickly transferred to a speedboat, then led roaring at over twenty knots toward Portsmouth with two Royal Navy destroyers nearby and soon a flight of Spitfires giving further cover overhead. He made it ashore, was given a fast motorcade ride to London, and, after a quick debriefing and delivery of the precious Würzburg parts, was granted as much leave as he wanted. By midnight he'd made it all the way to his home in the small town of Wisbech, in East Anglia. There was one fire in the house, and his father, mother, grandparents, wife, and toddler child were waiting for him around it. Cox marched in, and as he remembered:

" 'Hello, family,' I said, 'I've been in France, that's where I've been, and it's in the London newspapers tonight. How about that, then?' "

Cox was a hero, but what he'd brought over would be used for one of the largest killings in modern times. The Würzburg radar was even more advanced than British specialists had anticipated. It poured out waves of a bare ten inches from peak to peak. Those small, precise waves could be aimed with far more accuracy. The Chain Home system, with its yards-long waves, was antiquated by comparison.

That was bad enough, but the serial numbers Cox had captured allowed for a very disturbing calculation. Some of the replacement parts in the Würzburg had been installed in December, and the serial numbers on those parts were close together, suggesting that not many machines were in operation. But the replacement parts that had been installed in February

had serial numbers much further apart—which ominously suggested that many more spare parts were being produced, to
supply a much greater number of Würzburg machines now in
use. That explained the attacks the RAF was suffering. The multitudes of new Würzburgs, with their ability to spot incoming
aircraft, and their uncannily accurate guidance of searchlight
beams, were turning the sky over Europe into a killing zone for
Allied fliers.

But then—and could the full consequences of this discovery
ever have been averted?—the London analysts found a weakness in the Würzburg. Its settings were very hard to change, and
at first no one knew why. The clue came from the young operator that Cox had helped to capture.

He was very young and willing to talk. ("We spent the afternoon," Jones recalled, "sitting on the floor with him, fitting the
various pieces together, and listening to his comments.") But
he was also almost entirely ignorant of how radar worked. ("He
seemed to have spent more time inside prison than out of it.")
Britain had a lot of skilled amateur radio operators to draw
from, but in Germany it had long been strictly forbidden for
private civilians to build radios. Nor could German women be
recruited, even if they had an aptitude for technical jobs. In a
dictatorship obsessed with eugenics, they were to remain fertile,
demure, and at home. German radars had to be very simple,
Jones gradually realized, for there weren't enough educated
men available to repair and operate radars of any complexity. In
other words, Germany's most advanced radar had been built to
be an immensely sophisticated, idiot-proof piece of machinery.

It was the rigid nature of Germany's radar that made Jones and others in the Air Ministry believe they could turn Germany's technical advances back against them. For several months already, the RAF had been considering a weapon that at first seemed too simple to be of any great importance. It consisted merely of large numbers of aluminum strips, like long pieces of confetti, that could be dropped from an airplane. (The weapon's original code name was "Window," but it later became known as "chaff," the name we'll use.) If these strips were released from a fleet of planes, researchers believed they could act as a flurrying cloud, sending back an immense number of electric pulses. Germany's cities were increasingly dependent on radar for their defenses: there were radar–guided searchlights, radar–guided antiaircraft batteries, and—increasingly effective against British air attacks now—powerful, fast night–fighter planes, guided by information radioed up from still more radar sets on the ground. If chaff worked and the German operators were overwhelmed with false signals, that radar would be useless, and the attacking planes would be effectively invisible. Since British specialists now knew the Würzburg's exact wavelengths, they could use that to work out the ideal size for the chaff.

Watson Watt knew who was gunning to use chaff. It was Arthur Harris, the head of RAF Bomber Command. Harris had long known what chaff could do, but he'd held back, for chaff seemed to be the sort of weapon that could be used with full effectiveness only once. After a while the enemy surely would work out ways to distinguish the slow fluttering of the chaff from the faster speeding planes; they might also use chaff technology to jam British radar. There had been an impasse—a bit like everyone having poison gas but no one using it. But what the Bruneval raid had revealed was that Germany's radar sets— and especially the super-accurate Würzburg—were so rigidly calibrated that it would be hard for any of the operators to adapt to the new weapon. Chaff might render fleets of planes impregnable for a long time. Watson Watt knew he had the battle of a lifetime coming up. So did Harris—and he was sure that, this time, he was going to win.

It's doubtful that there was a more disagreeable character on the Allied military side in World War II than Harris. He could be kind to his immediate family, but he had few friends and no hobbies. He never read a book, and he never listened to music. He had only one great passion in his life, and it was a hatred. It wasn't directed against Germany. It seems—from the evidence of his actions—that it was directed against blue-collar workers.

Harris was an extreme reactionary. Like many well-off individuals of his time, he often expressed great distaste for the British working classes, and for their German counterparts as well. The writings of even many ostensibly gentle literary intellectuals from this period are disturbing when it comes to this topic; indeed, they bear some resemblance to the racial hatred that Japan and America came to feel toward each other as their battles in the Pacific went on. For American military leaders, this led to the burning down of entire Japanese cities, with few moral qualms; for Harris, it led to a cold and pitiless view of any workers or children who on the ground would be forgotten when his bombers came overhead.

Many officers aware of his plans were appalled at what he wanted to do. The United States, for example, was bombing enemy factories, railroads, and docks. Often there were vast mistakes, and civilians were killed. But at least in Europe that was never the goal of entire campaigns, and American flight officers who consistently missed their navigation targets could be removed from duty. The Royal Navy also wanted to use whatever bombers were available to concentrate on submarine factories and shipyards, and, if possible, to use bombers to target enemy submarines or surface ships on the seas.

Harris saw it differently. Enemy factories may have been his ostensible primary targets, but he was convinced it was a waste of time to try aiming precisely at the factories or construction yards. Nor did he want his planes circling aimlessly over the sea on the hunt for enemy submarines. That was merely a distraction. He wouldn't allow it, and if it had to happen, he wouldn't encourage it. He wanted to kill people, quite as much as he wanted to destroy buildings or equipment. The huge supplies of high explosives and incendiaries that the RAF was accumulating were to be dropped on the workers themselves, in the houses where they lived. That, he was convinced, was the most effective way to destroy the enemy's power. In the very month of the Bruneval raid, Bomber Command released Directive 22, which was meant to ensure that in all attacks "[the] aiming points [are] to be built-up areas, not, for instance, the dockyards or aircraft factories. . . . This must be made quite clear."

Objections were regularly brushed aside, whether they came from military officers or expert civilians. The Bruneval data now helped give Harris the argument he could use to take his efforts to the highest level. He would use chaff to turn off a city's radar protection. Then, when the city was defenseless, he would do as he liked: trying to destroy its factories, perhaps, but also destroying whoever lived there.

Watson Watt was frantic. This was never what he'd devised radar for, but he was just an underling now, and despite a last desperate rush of words and memos, he could merely watch as the remarkable defensive weapon he'd helped create was lifted from his control. He even tried getting Henry Tizard to help him. Tizard was the man who'd headed the original committee that created Britain's radar system, and that had been so crucial in the 1940 Battle of Britain. Tizard also despised Harris, and now he started building alliances that in normal times might have been enough to stop him. But everything had to get past the man Tizard had humiliated at the radar committee in 1936—and Lindemann had the exclusive ear of the prime minister now. It was with the greatest pleasure that Lindemann ensured nothing Tizard proposed was seriously considered by the government.

By early 1943, Tizard and Watson Watt knew they had lost. At one point Harris sponsored a talk at Bomber Command's Buckinghamshire HQ on the Ethics of Bombing. After the talk, the

Bomber Command chaplain Rev. John Collins stood up and said that, on the contrary, this was the Bombing of Ethics. But he was firmly corrected, and no one else there dared to speak in his support.

There was little question what city Harris would select to show what his force could do. Hamburg was a huge industrial center, with lots of densely concentrated worker housing. It was also on the North Sea, with the river Elbe going through it. Navigating at the border where land and sea come together is especially easy (for land and water respond very differently to radar, as we'll see).

Harris gave instructions to guide his RAF navigators as they entered airspace over the city. To the south of the Elbe were the factories and famous U-boat construction yards: Blohm & Voss, Stülcken, and Howaldtswerke. Those were the targets that the Royal Navy and Tizard wanted destroyed. But Harris's pilots had express orders to avoid them and stay on the *north* side of the city. There were no war factories there, only row after row of tenements in six-story blocks much like London's old East End or parts of New York's Lower East Side. Some of the men there were employed at the factories, but the majority of inhabitants were older people, women, and—since there had been only partial evacuation to the countryside—lots and lots of children.

The Bruneval data was processed by that spring, and small, downward-aiming radars were in place in many of the aircraft; the very final arguments for chaff were won by early summer. Now all that remained was to wait for the ideal weather conditions. In Hamburg, July was a warm month, with temperatures in the eighties. The humidity stayed unusually low for several days. Harris monitored the weather reports.

There were a number of powerful preliminary raids, but it was only on the evening of July 27 that the main RAF force set out. A few couples were still strolling in Hamburg's parks and open areas; it would be hours before any signs of the RAF appeared.

At eleven P.M. the planes were far over the dark North Sea, still unseen. In the radio headsets of the aircrew, broadcasts sent rippling out from England led to electrons vibrating over minuscule distances, and then magnified into audible sound. Inside the downward-looking radars, more electrons were hurtling through their miniature copper-bound channels, flinging down tight radio waves. There are relatively few loose electrons in water to respond, so when the oscilloscopes showed only blackness, the navigators knew they were over the cold waters of the North Sea.

But then, about an hour later, the invisible waves pumping down from the onboard radars began to hit something different. Metal structures such as sheds or rails have great numbers of loose electrons; tree leaves and brick buildings and paved roads have fewer, but transmit some signals as well. The oscilloscopes two miles high began to reveal a sharp contrast with the blackness of the sea. The Pathfinder planes at the front realized they were crossing the coast and made slight corrections to be exactly on course. More than seven hundred bombers followed close behind.

The bombing crews started expelling the confetti-like bundles of chaff from the planes. The bundles blew apart in the fast-moving air, separating into thousands of aluminum strips as they fell. The invisible waves that the Würzburg and other sets were spraying upward slammed into all the loose electrons in the slowly fluttering aluminum. As the aluminum's loose, outer electrons moved back and forth under that unstoppable power from below, they became miniature transmitters. The sky was still pitch black to human eyes, but for each radar set below it glowed bright as each aluminum strip began broadcasting. Millions of identical signals rained down.

The Würzburg and other radars were swamped. No ground controller could pick out the actual airplanes in that dazzling electric power. The searchlights they had controlled suddenly began spinning aimlessly; antiaircraft batteries either stopped operating or fired randomly. Fighter pilots called down frantically for directions. Some of the ground controllers in such circumstances yelled into their radios for the fighters to "Break off, the bombers are multiplying themselves!" Others sent angry radio messages up, telling the fighter pilots to fly twisting flight paths in an attempt to have them stand out from the multiplying aluminum transmissions that filled the sky.

None of it worked: the RAF planes were now unopposed. High explosives were dropped first, to pierce the water mains (a later count showed they had been broken in over two thousand places) and to break open the houses below. Bricks were shattered and fragments blew apart. Then the main bomb doors opened, releasing the chemical incendiaries.

Much of Hamburg was built of wood, and wood is made

when the miniature photovoltaic units we call leaves take ordinary, separate carbon atoms and hook them together in long chains. It takes months or even years' worth of light energy pouring down from the sun to hook up carbon atoms that way.

When the RAF bombs shattered the clusters in the wood, each carbon atom was on its own. By itself, that would have meant a great deal of rubble and dust, and many people hurt from the collapsing wood, but the damage would then have been over. Yet it didn't end there, because the explosives the RAF had deposited sent out immense amounts of heat.

The heat hurtled along the Hamburg streets, transforming everything in its path. It soaked into dust flecks in the air till they exploded, and it heated the carbon in Hamburg's wooden homes so much that they reacted with oxygen and exploded into flames as well. The energy that the sun had poured into that wood over the long years when it had grown in forests now reappeared, in a sudden horrifying burst.

We can't see the electric waves of radar, but in the fury of the burning buildings, the electrical waves coming out were shorter, more intense. When they struck human eyes, retinal cells sent signals to the brain.

In the Hamburg firestorm, Faraday's invisible waves were turned into light.

Fires began, and the flames joined together, and then the whole city ignited. People tried to escape, but how? One fifteen-year old girl remembered: "Mother wrapped me in wet sheets, kissed me, and said: 'Run!' I hesitated at the door . . . but then I ran out to the street. . . . I never saw her again."

An older girl, nineteen, joined a group trying to get across the large boulevard of the Eiffestrasse, but at the last moment she realized she had to stop. The heat of the fires was making the street melt:

"There were people on the roadway . . . alive but stuck on the asphalt. They must have rushed on to the roadway without thinking. Their feet had got stuck and they had put out their hands to get out again. They were on their hands and knees, screaming."

Up above, a captain in one of the Pathfinder squadrons—at age twenty-seven, older than almost every other pilot—looked down at the storm that had been a living city. "Those poor bastards," he muttered over the radio. He pushed his hands on the steering control, and the great plane began to turn. Insulated wires in his cockpit guided electrons along copper strands as indications of the wing's dipping appeared on his cockpit displays; Faraday's waves flooded in through his thick glass windshield, some broadcast invisibly from the thousands of fluttering aluminum strips in the air; others broadcast visibly, painfully so, from the glaring flames. A final glance, and his Lancaster bomber swung away. The single night's attack was over—but the bombing would continue on and off for another two years.

The attack was shattering, yet all that horror and struggle—the Chain Homes defenses, as well as the Hamburg-destroying war machines—only skims the surface of what electrical effects can produce. For there's yet *another* level that goes beyond the image of powerful waiting electric charges, beyond even the invisible, space–crossing waves that can force those charges to move. The vision that Maxwell had of atoms was incomplete.

In the 1910s and 1920s—even before Watson Watt found himself in Slough—a small number of theorists had begun exploring this new, submicroscopic world. If they were right, then the world beneath our own is composed of electrons that travel in abrupt teleporting jumps—known as "quantum" jumps—and also in sudden stops and starts.

This would change everything, for electrons are central to electricity, and whenever we discover something new about them, the groundwork is laid for a fresh technology. In late Victorian times, the vision of electrons as hard little balls led to the technology of telephones, lightbulbs, and electric motors. Faraday's and Hertz's understanding of waves had led to radio and radar, which were so central to World War II. Now the realization that electrons could dematerialize—that they could in effect pop through space and be made to start and stop in fresh places—would open the way for yet another device, a thinking machine, which would shape our era as much as electric lighting and telephones had shaped the nineteenth century.

In the 1920s the English word *computer* (and its cognate in other languages), still meant a person, usually female, who spent laborious hours at a desk, using a mechanical calculator, or even old–fashioned pencil and paper, to compute whatever dull arithmetical task was assigned. It seemed impossible to go further, for if any genuine thinking machine were to match the quick flicks of human thought, it would need to shift its internal circuitry far faster than anyone could then imagine. No solid, mechanical object could do that.

But perhaps the wildly teleporting flights of tiny electrons could.

PART IV

A COMPUTER
BUILT OF ROCK

*T*he metals that were mined for the planet's war machines carried electrons that could leap instantly from one adjacent atom to another, never appearing in the space in between. But in other substances those leaps did not always so easily take place.

When those clusters of atoms came together, in common rocks and crystals scattered across the planet's surface, the electrons powerfully blocked one another's flights. Newly arrived electrons might try to move faster—to increase their energy levels—but an exclusion zone seemed to operate, keeping them at bay. The electrons in these common rocks and clays were slowed; they were almost stopped.

Humans had spent a century transforming their civilization by making fast electrons serve their purposes. Now the twentieth century's bloodiest war had ended. The power of slow electrons was still to be unleashed.

9

Turing

CAMBRIDGE, 1936, AND BLETCHLEY PARK, 1942

There had been some efforts to build a computer in 1820s England, but the prevailing technology of steam engines and ball bearings and metal cogs was too crude ever to make it work. The failure was not just in technology but in imagination. Even a full century later, in the 1920s, there were many ingenious machines in the world—there were locomotives, and assembly lines, and telephones, and airplanes. But each did only one thing. Everyone accepted the idea that to get a different task done, you needed to build a different machine.

Everyone was wrong. Alan Turing was the man who first showed in persuasive detail how it would be possible to change that. His life ended in tragedy, for although he conceived a perfect, clearly describable computer, and although the new insights about how electrons can leap or seemingly stop might have allowed him to construct it, the technology remained elusive. New ideas in science don't automatically produce new machines. He would be lauded in death—but not while he lived.

As a boy, in the 1910s and early 1920s, Alan Turing loved the way he could think his way out of problems. He had trouble distinguishing right and left, so he dabbed a red dot on his left thumb, and then was proud that he could get around as well as other children his age. Soon he could outnavigate both children and adults. At a picnic in Scotland, to get his father's approval for being suitably brave and adventurous, he found wild honey for the family by drawing the vector lines along which nearby honeybees were flying, and charting their intersection to find the hive.

But as an adolescent and then a young adult, he found it harder and harder to blend in. By the time he was sixteen he realized that he was physically attracted to men, which was bad enough, but he also realized he was without question an intellectual, and in 1920s England, especially at its private schools, that was even worse.

His father was far enough away, serving in India with the Civil Service, not to have to pay much attention, but his mother, who was from a proper upper-middle-class background, would have none of it. Alan was a normal boy, she insisted, who would one day learn to control his strange musings on beauty, consciousness, and, above all, on science. She was sure he would also—as he seems to have dutifully suggested in his letters from prep school—quite soon bring back for a visit one of those pretty girls he hoped to meet at nice London parties.

Instead, by age seventeen, he'd fallen in love with an older boy at his school, Christopher Morcom. They built telescopes and peered out of their dormitory windows late at night. They read physics books together, and talked about stars, mortality, quantum mechanics, and free will. In their discussions, they "usually didn't agree," Alan happily wrote, "which made things much more interesting."

But then, just a few months after they met, Morcom died of tuberculosis. Turing had been reserved with his mother until then, but now opened his heart: He and Morcom had always felt there was "some work for us to do together," he wrote, ". . . [Now] I am left to do it alone." But what was that work? Many people question their faith after someone they love dies, but adolescent deaths are raw, immensely so: the survivor experiences the intensity of adult emotions, yet can't place what happened in familiar cycles of life. A hole is ripped in the universe.

Turing seems to have lost whatever religious faith he once had. He angrily dropped the usual Edwardian belief that only the body is lost in death, and that an immortal soul, not made of any earthly substance, lives on. Morcom was gone. People who tried to comfort him by saying his friend somehow survived were liars.

That anger, that belief in cold materialism, was indispensable for the great electrical device that Turing imagined just a few years later. It's hard to conceive of creating an artificial device that duplicates human thinking, if you believe in an immortal soul. The perishable stuff that the computer has to be made of—the wires or electrons or whatever—will lack all semblance of that soul. But if you're sure, with all the anger of adolescence, that nothing but dead earth is what remains when we die, then cold wires will do just as well as any living being.

For several years Turing acted the part of a contented Cambridge undergraduate, but he often turned to the Morcom family when he was stressed, either through visits or heartfelt letters to Christopher's mother. Eventually, to his Cambridge friends' puzzlement, he began repeatedly quoting a line from Disney's new *Sleeping Beauty* movie, about the poisoned apple and how quickly biting it could produce eternal rest.

Early in the summer of 1935, when Turing was twenty-two, he came across the problem that triggered his major work. It dated back to the turn of that century, when, at a lecture hall in Paris, on a hot August day, the great German mathematician David Hilbert had read out loud what he considered the most important mathematical problems for the twentieth century. A follow-up of one of the hardest—still unsolved when Turing heard of it—dealt with a deep problem in logic, asking how certain long chains of reasoning could be carried out. Most researchers assumed the answer would come with an abstract mathematical proof. Turing, however, had always liked to tinker: he was excellent at building radios, fixing bicycles, and putting together metal contraptions of almost any sort. Now, as he lay in a meadow in a small town near Cambridge after one of his long, solitary afternoon runs, he imagined an actual *machine* that could crank through the steps of Hilbert's logical problem.

In the next few months, Turing showed that this imaginary machine could resolve the questions Hilbert posed about how to prove the truth or falsity of any abstract statement. The machine would need electricity, of course, perhaps in a form not yet imagined, but that didn't preoccupy Turing now. Instead, he wondered what *else* such a perfect machine could do. This took longer, for he realized that, in theory, a machine that clicked through these strings of logic could quite likely do almost anything.

All that the machine's operator would have to do was write down, very clearly, the instructions he wanted it to follow. The machine wouldn't have to understand what those instructions meant; it would simply have to execute them. Turing proved that almost any action he could imagine—adding numbers or drawing a picture—could be translated into simple logical steps that a machine could follow.

If a critic protested that the machine wasn't as powerful as Turing proposed, and named some other tasks that it couldn't perform, Turing would simply have the critic break the tasks down into discrete steps, and describe the steps using this same clear, logical language. Then Turing could give those instructions to the machine and it would chug along, faithfully carrying them out—thus showing that the critic had been wrong. We today are so used to machines carrying out sequences of instructions—we automatically assume that a computer or cell phone will follow our tapped-in commands—that it's hard to remember a time before it was accepted. But when Turing was a student hardly anyone imagined that inert machinery could accomplish such intelligent work.

It was a stunning intellectual achievement, but also a lonesome one. This "Universal Machine," which Turing described in his 1937 paper for the *Proceedings of the London Mathematical Society*, was self-contained and entirely without emotion. If it was fed the right instructions, then from that point on, it could operate on its own—forever.

The machine wouldn't even need an operator to reach inside and change it when the tasks it faced changed. For Turing had also begun developing the concept of software. He realized his machine wouldn't be useful if it had to be rebuilt each time it was given a new problem to work on. Instead he imagined that the inner parts of the machine could simply be rearranged as needed. This software might seem to be a part of the solid substance of the computer, but it would really be constantly shifting around, configuring itself in one restless way, then another.

This is where electricity comes in. Turing's imagined computer couldn't simply have a lot of wires locked into one particular configuration. For when we think, we're comparing and combining lots of different sensations and thoughts; we perform a huge number of rearrangements with them, and we do it very quickly. If Turing's computer were to match the human mind, it too would need lots of switches that could rearrange themselves just as quickly. The switches would have to be so small and work so fast that miniature metal cogs and gears—the stuff of conventional adding machines—wouldn't suffice.

Telephone companies had reached the limits of simple metal switches a long time before. Their first improved switches consisted of sturdy young men who physically took out plugs from one line that ran up through a hole in a large board and pushed them into another line. (Since that board was where all the switching took place, the term *switchboard* was soon born.) When supervisors found that the men swore too much (and also that they were easily roused to join unions), the males were replaced by more genteel females; when Bell System administrators in the late 1890s found that even great offices filled with such women were overburdened, the first semi–electrical switches came into use.

Those switches used very thin metal wires that acted like miniature swiveling drawbridges. Electrons poured up to the mouth of the drawbridge, and if the bridge was in position, the electrons would hurry across the gap. But if the drawbridge wire had been raised or had swiveled to one side, the electrons skidded to a halt, or just fell aimlessly into the gap, and the signal they carried didn't get through.

Unfortunately, even the most advanced telephone switches of the 1930s were far too big for Turing's purposes. The electric thinking machine Turing envisaged would be sorting and arranging so many different "thoughts" that it would need thousands, perhaps millions, of simultaneously operating switches. That many metal wires, however slender, wouldn't fit.

What he needed, of course, were the new insights in physics about "teleporting" electrons and the rest of what was termed quantum mechanics. The new theories suggested the electrons could do the switching without having to be guided along slowly swiveling wires. Instead, if the right quantum rules could ever be applied, electrons could be made to jump and switch position even within solid, unmoving matter.

That was the dream, and Turing had studied enough physics

to know of the new research in quantum mechanics. Many of the founders of that field were working around him in Cambridge. But, like other engineers and mathematicians, he believed that quantum effects were perpetually hidden away from us, limited to a submicroscopic realm too small ever to be of use. It seems, at this time, he never seriously considered that quantum switches could be his answer.

He still missed Morcom, but knew he needed a new kindred spirit, someone to share his increasingly intense thoughts about consciousness and machines—perhaps even about what it meant if we were just flitting arrangements of software, yoked to Yeats's dying animal. Because of the professional fame from his Hilbert papers, Turing was invited to spend some time at Princeton, where the equally brilliant Professor John von Neumann taught. At first von Neumann seemed to be the intellectual companion Turing sought. But von Neumann had always been a great chameleon. In Hungary, where he'd been raised, his first name had been Janos, then in Göttingen he'd happily become Johann, and now in Princeton he was good old Johnny. He'd become hyper-Americanized, and insisted on a social life that was dominated by loud cocktail parties. Turing seems to have gone to a few of these parties, but he could never reach von Neumann emotionally, and returned to England after only a few semesters.

During World War II, Turing was hired as part of the British government's codebreaking group at Bletchley Park, in the south of England. There were academics there, exulting at being freed from the stuffiness of old Oxbridge College, and there were crossword experts and naval officers and a director of research for the John Lewis department stores (who was also a chess champion), and in this mix Turing came alive. He'd always liked using his wits for practical goals—at Cambridge he'd casually taught himself to tell the time by looking at the stars—and at Bletchley, around the impromptu cricket pitches and the red-brick huts on the manor house's grounds, he got his chance. He was assigned to the team working on cracking the Enigma machines, which much of the German army and navy used to encode their messages. Within a few weeks Turing had helped devise new techniques for codebreaking, and within two months he was head of the unit breaking all German naval codes.

In late 1940 this was one of the most vital of all British operations, for if more of the U-boats that were destroying Atlantic convoys could not be found, Britain would slowly but inevitably starve. His unit needed to build some sort of device that could at least partly duplicate the inner workings of the German machines and churn through tens of thousands of permutations each hour. The British device, running with cogs and punch tape and simple circuitry, was given the label "bombe."

It wasn't a computer, for lots of clerks were needed to run the bombe devices, first dozens of them, then hundreds. Almost all were young women, recruited usually from proper upper-middle-class families into the Women's Royal Naval Service—Wrens, for short. But there was something curious about this combination of humans plus machines. For with the Wrens and the bombe device working together, Turing had created the closest thing yet to the universal computer he had imagined in the quiet prewar years. That combination of "Wrens plus the simple electric circuits in the machines they supervised" was, in a sense, the "hardware." When fresh data came in from the outside world—when the Germans switched to a different encoding procedure, or a colleague proposed a new approach—Turing simply had to "reprogram" the whole group working for him.

Progress was slow at first. Sometimes convoys sailed toward areas where the Admiralty knew there were packs of U-boats waiting, but the clacking Wren-directed machinery couldn't break the intercepted German shortwave signals in time. They would realize with dread that ships were exploding, sailors drowning. But Turing was a master at organizing small groups, and increasingly often they succeeded. His "computer" didn't yet exist as one centralized physical object, but he now had the equivalent of most of its parts—the memory, the processor, the reconfigurable software—working away, albeit scattered in separate buildings, and made of such disparate elements as female Wrens, copper electrical wires, and Turing's own thoughts.

And then he fell in love—or at least very much "in like." Joan Clarke was a young Cambridge mathematics student who'd been sent to his division at Bletchley. He told her that he had "homosexual tendencies," but she didn't seem to mind, and they began the sort of romance that any overworked, mathematically fascinated young couple would find normal. In their free moments they would lie on the Bletchley Park lawn together, commenting on the mathematical patterns in the daisies they plucked for each other. At dawn, after exhausting nine-hour night shifts, they played bouts of what Turing called "sleepy chess" (using clay pieces he baked on the hot-water radiator in his lodgings). Once Turing knitted a pair of gloves, but wasn't dexterous enough to finish off the fingertips; Joan, only slightly teasingly, did it for him.

The greatest electrical progress began in February 1942, when the German navy shifted to new forms of coding. All the advance intelligence that Bletchley had been providing on U-boat wolf packs suddenly disappeared. Yet airborne radar wasn't quite ready to take up the slack. The Royal Navy, and its new American allies, were operating blind. Turing was tense; he picked at the sides of his fingers so much that an open sore formed. The improvement that the German navy made in their Enigma machines was something Turing eventually cracked, using the technology of his bombe devices. But at the same time German forces had developed a new cipher, to use in such top communications as High Command messages, and this the simple bombe couldn't penetrate.

From this impasse came funding for others at Bletchley to build a far more advanced machine called Colossus, one that was several steps closer to Turing's 1930s vision. Once it was working, its hundreds of miles of interior cables generated so much heat, apparently, that the Wrens who adjusted its plug settings sometimes had to ask all men to leave the premises, so they could undress enough to work in the heat.

Turing's work had been important enough that he was awarded the coveted Order of the British Empire. A memo he wrote to Downing Street about delays in supplies to Bletchley had even resulted in Churchill jolting his principal staff officer with a memo: *"ACTION THIS DAY: Make sure they have all they want on extreme priority. . . ."* But even so, the Colossus still needed to be laboriously reconfigured almost every day, as the Germans regularly changed the details of how messages were sent. The machine was more than a calculator, but not yet a true programmable computer. It could make only the simplest of sorting decisions on its own. Compared to what Turing had imagined, it was still excruciatingly slow.

When the war ended, in 1945, Turing was eager to go further. On one of his last days at Bletchley, as he and Joan filed away their hyper-classified work, Turing told her that he'd persuaded the National Physical Laboratory (NPL) outside London to give him a team so that he could, finally, build the infinitely changeable machine he sought. He and Joan had broken up—on calm reflection, she'd decided marriage to a gay man would not be the wisest of choices—but this was a moment to share confidences.

At first it went well. The head of the NPL was Sir Charles Darwin, grandson of the great biologist, and one of the best-connected science administrators in the country. What Turing didn't realize at first—though he learned fast—was that although Darwin had been a useful scientist as a young man, and indeed had helped Rutherford with the original solar system model of the atom, he was now a harrumphing, pipe-smoking grandee.

Darwin seemed to feel that if this odd fellow Turing wanted to build a practical machine, or even a number of practical machines, that was fine. After all, postwar Britain needed such help. But why was he going on about this "universal" machine, and what were all those musings he heard Turing blurt out—it had been difficult keeping him away from journalists—about a computer that could someday write a sonata, or be given mechanical legs and navigate the countryside, or be programmed to learn anything one could imagine?

They had meetings, and when Turing spoke about the need for vacuum tubes and telephone switching relays, Darwin was interested. When Turing spoke about pure math, he was polite. But when Turing insisted on speaking about software, and suggested that they didn't *need* to build a complicated machine, that their goal was to make the machine "do all kinds of different things, simply by programming rather than by the addition of extra apparatus," Sir Charles decided that this young man did not understand the modern world at all. How could one possibly build a machine by, expressly, not building a particular machine at all? There were rumors that Turing had done something very important in the war, but to Darwin he clearly seemed off the rails.

Turing was exasperated. The only way to get a computer to operate, he knew, was to create a physical machine that was quite simple, with an arrangement of hardware inside so pliable that one could devote one's efforts to coming up with ingenious *programming* that would change how the electrical signals inside were shuttled around. There was no need to build the physical machine afresh each time. His fellow workers had come close to this design with the wartime Colossus, in their efforts to keep up with the constantly changing arrangements inside the German's coding machines. At Bletchley there hadn't been enough time to do more. Why wouldn't anyone let him go further now?

It didn't help that Turing still had no sufficiently miniaturized components for his switches or his storage devices. (The software inside his machines could just be a changeable pattern of electric circuits, but there would also have to be solid hardware, such as a memory section to record what the software came up with.) Yet the only components that were available were large and bulky. To help with the memory devices he tried a trick from the radar project of World War II, where technicians had found that if they filled a big round pipe with gallons of dense liquid mercury, and then sent pulses of liquid waves through that mercury, those pulses bounced back and forth inside the pipe with surprising accuracy.

The pulses seem fast to us, but are so slow on the time scale of electrons that the pattern lasts long enough to form the desired memory device. Yet funding was so poor that at one point Turing was reduced to scavenging the suburban fields outside the NPL labs for fragments of plumber's pipes to use in his grand computer's memory. As 1946 turned into 1947, Turing made hardly any progress. Darwin and the other administrators became suspicious. Turing looked for other jobs.

In the autumn of 1948, Turing moved to Manchester, where word had it that something closer to a true computer was being built. (There were competing projects in the U.K. now, for a number of researchers had picked up on Turing's prewar papers, or on American wartime work similar to the still-classified

Colossus machine. All the teams wanted to take this research further.) But if London and Princeton had been difficult for Turing, Manchester was impossible. The mathematicians designing the device were friendly enough, but they had already started on their own blueprints, which weren't even as bold as Turing's original 1937 vision.

The practical staff in the engineering labs might have been able to modify that design, but they were wary when Turing, with his prep-school, south-of-England accent, introduced himself and tried to get them to help. Everyone at Bletchley had worked together, but that was then. The unity of wartime was fading. The Manchester engineers had a lifetime of experience to show that someone with an accent like his was likely to ignore them smugly, or at least not have any useful advice to offer. Turing knew that he was different. His years with radios and punch-card readers and building electrical systems of almost every sort would have allowed him to give them lots of useful pointers. But with the class system so firmly, invisibly, in place, it was impossible to convince them. Turing was stuck.

In fact there was a new technology being perfected in America that could have transformed his life—one that involved harnessing the newfound properties of electrons in the submicroscopic quantum realm. Turing had even heard enticing rumors of these devices. But the silent class war meant he couldn't work with the engineers to chase it down, and anyway, there was no sign of it actually being produced.

He dawdled for most of 1949, trying one subject after another. At one point he considered trailing back to Cambridge, but becoming an academic seemed petty after all he had done. He tried some of his old topics in pure mathematics, but he was getting too old for that as well. Von Neumann wrote him a cheery note from Princeton ("Dear Alan, . . . what are the problems on which you are working now, and what is your program for the immediate future?"). But Turing had little to answer. He was alone, perhaps more so than ever before.

For a while he went back to analyzing the patterns of swirls in daisies and other plants, perhaps remembering those warm days on the lawn with Joan Clarke. He also began to think about what it really meant to be a solitary being—and from that he came up with a paper on artificial intelligence and the nature of self-awareness, which just a few years after his death would be recognized as fundamental to modern cognitive and computer studies. (His ideas on computer simulation of biological development also became important for today's research.) But in postwar Manchester, where only fragments of the ideal computer he'd conceived were being built, daisies and isolated minds were all just a further sign of this strange man's irrelevance. Turing was shunted aside once again. Perhaps he accepted it.

His mother continued to write him regularly, apologetically asking about his progress in finding a wife. It was getting harder to write back with the conventional lies. His love life was meager, and perhaps in memory of the cold showers from prep-school days, Turing had taken up long-distance running, and was at one point ranked as one of the leading marathon runners in Britain. (His best time was within seventeen minutes of the then-current Olympic record. He would have been considered for the British Olympic team but for an unexpected hip injury.)

Instead of real love he was resigned to occasional casual pickups. They were of no great significance until, one evening in January 1952, Turing realized that a young laborer who had stayed over must have passed information about his house to an accomplice, for when Turing arrived home, he saw he'd been burgled. He went to the police to complain, probably more hurt by the breach of faith than by the cash value of what was taken.

It was a terrible mistake. Homosexuality was a crime in Britain then. Possibly in Cambridge it might have been handled with a reprimand; in London not long after, the actor John Gielgud would be arrested for a similar offense, yet lobbying by his friends allowed him to minimize time in jail, and newspaper reports were not drawn out. But Manchester was neither London nor Cambridge. The accomplice was arrested, immediately implicated Turing in exchange for immunity, of course, and Turing was seized. He was soon in court, alone, for what was then considered a very serious crime.

Because of his wartime service—the award from the British government; hints that he might be selected for a knighthood—it was arranged that he could avoid prison. But this meant agreeing to undergo experimental treatment to "cure" his homosexuality. The treatment was mandatory consumption of female hormones. There was little alternative to accepting. Prison would have been brutal, and only at a university could Turing continue his work.

Turing began the course of pills, picking them up at regular intervals as required. At first he thought the effects would be insignificant, but the treatment made it hard to concentrate. Even if he had tried to get the dosages lowered, the judge was implacable and wouldn't have allowed it. And then, as the hormone treatment went on, Turing found to his horror that he was developing breasts.

It was too much. He had no intellectual companions, he had no chance of love—and now his body and mind were being destroyed. The treatment stopped in April 1953, but he never really recovered. One rainy June evening in 1954, Turing was at his home in the Manchester suburb of Wilmslow. He took out an apple, and then he opened a jar of potassium cyanide that he kept for gold-plating reactions in his electrical experiments. The jar still had some cyanide inside when it was examined the next day; the apple, found beside his body, was missing several bites.

10

Turing's Legacy

NEW JERSEY, 1947

What Turing was hunting for—which might have saved his life—was right under his nose all the time. He'd even received a hint in 1948, even before he moved up to Manchester. A friend from Bletchley days, Jack Good, had written him:

"Have you heard of the TRANSISTOR (or Transistor)? It is a small crystal alleged to perform 'nearly all the functions of a vacuum tube.' It might easily be the biggest thing since the war. Is England going to get a look-in?"

But then there was only silence. Something had happened in America to slow down the new device's development, and the reason goes to the heart of how it worked.

When Turing was a student, most electricity specialists felt that all the substances in the world were divided into two quite different types. There were substances such as steel or copper that could transmit an electric current, and there were substances such as glass or wood that could never conduct a current. The first group were called conductors, and the second were insulators, They had as much in common as aardvarks and coal mines: the two categories just did not overlap.

That simple distinction seemed to explain the age-old question of why you can see through glass but not through steel. The inside of a steel wall looks a bit like a huge, abandoned Egyptian temple, full of neatly aligned pillars of iron and carbon atoms. Seen in close-up, however, those atoms aren't smooth or neat. Many have lost their outermost electrons entirely, and—as we saw in the first radar chapter—those electrons float aimlessly inside the steel. When light flies in, it steadily gets used up as it shoves those electrons so that they start moving with greater energy. This means that as a light wave goes deeper into the steel, there's less of it that hasn't been diverted—"soaked up"— by those lurking free electrons. It's as if a wave of explorers entered the abandoned temple and, one by one, were dragged behind the pillars. Pretty soon there aren't many explorers left. Light waves reflected from you can go in one side, but won't make it out the other.

Inside a wall of glass, by contrast, the atoms are better behaved. Their outermost electrons are much more tightly bound to their atoms and have no interest in waylaying explorers. A light wave that enters their domain will soar in and out unscathed, emerging as bright as ever. Light waves reflected from you will travel all the way through—you can be seen by someone standing on the other side.

That difference is why metals conduct electricity, and glass doesn't; indeed, it is why electric wires often sit on glass insulators. A current hums easily through the copper or aluminum in those wires, because the metals have so many free electrons inside. The pushing force, the invisible whirlwind from the power station, just nabs and pushes them along. The current doesn't pass across the glass insulators, however, for glass doesn't have free electrons inside to become a moving current. A live electricity pylon is like a very, very stupid switch: it's always in the "on" position, for electric current just runs forward, along the wire, and never switches direction to go downward through the glass insulators the wire rests on.

If there were only these two possibilities—that some materials can always conduct electric current, and others can never do that—then Turing's legacy would consist only of a few interesting papers and some huge, constantly overheating rooms filled with plugs and vacuum tubes in complex arrangements. The computers we take for granted today wouldn't exist. But metal and glass aren't the only substances that exist.

Our universe has something else—and this gives rise to a third possibility.

When Turing went on his long runs, he often passed along hilly country paths, or even raced directly on the sandy beaches with which Britain, being an island, is so well endowed. Those sands and hills are composed in large part of the element silicon, as indeed is much of the surface of our planet—Mount Everest is largely made of silicon.

Radio technicians had been irritated by silicon for a long time. It didn't fit into the two accepted categories like everything else: it wasn't a metal that always conducted electricity, nor was it like glass or diamond, which don't conduct electricity. It was different, confusing. Much of the time, if a fragment of silicon was in a circuit it would seem to act like an ordinary, run-of-the-mill insulator. That was fine. If you led a wire into silicon, then any current traveling through the wire would skid to a halt when it reached that barrier.

What bothered radio technicians was that silicon didn't *always* act like this. Sometimes a chunk of silicon that you thought would be an insulator would somehow undergo what seemed to be an unseen twisting and stretching inside, and then it would no longer be a good, dependable, happy-to-stay-at-home insulator. Instead it would now act like a conductor, carrying floods of electrons along. It was neither one nor the other. It was a "*semi*conductor."

Silicon was so fickle that when the great research division at Bell Labs began efforts to create nonmechanical switches, one of its first directives—as wise as Disney firing Jeffrey Katzenberg just before he produced *Shrek*—was to cancel all research on silicon. Luckily, Bell Labs was also very big, and directives in large companies are easy to sidestep. There was at least one researcher there, Russell Ohl, who'd been intrigued by silicon's changeable nature for years. He put fragments of it in the circuits of radio receivers, which he then placed in his son's baby carriage. Then he'd stroll around New York, getting the pleasure of measuring exactly when silicon conducted electricity or not, while also giving his son a good airing. He was enthusiastic about the prospect that this sensitive, fickle substance could someday be useful. After Ohl's son outgrew the carriage, their happy silicon excursions were out, but Ohl still wanted to take silicon research further in his labs. When Bell tried to cancel his research, he finagled a way around the company's directives and kept a silicon group on.

By 1946 and then 1947, what was happening inside silicon was finally becoming clear, due to that early work of Ohl and others. Silicon will sometimes form perfect crystal lattices, stretching like a dizzying M. C. Escher drawing, where a three-dimensional scaffold stretches on to infinity. But perfection is hard to find on our planet. When silicon is found in the wild, or melted and then cooled down in a factory, it's more likely that tiny cracks and gaps will appear in the perfect scaffold. A few intruder atoms such as phosphorus readily slide into cracks in the lattice, and they bring intruder electrons. Nice, vulnerable *extra* electrons.

If the only thing those electrons did was travel along swiftly, the silicon would simply have become another switch that was always "on." But Ohl and the others knew enough of quantum mechanics to realize that the electrons which shared the spaces inside a silicon lattice could affect and sometimes slow each other, even over distances that on the atomic scale were huge. When just the right number of intruder electrons were put in there, these strange effects could be adjusted to suddenly make it impossible for the electrons to leap along in the way that would transmit an electric current. But with a slightly different adjustment from outside, produced by Bell researchers peering down at their lab-bench creations, poking and prodding and adding different substances or applying—ever so carefully— different force fields, the bizarre "slowdown" effects would end, and the electrons could be let free to quickly speed forward again.

The chemistry was too hard for any single individual to

handle, and Ohl was getting out of his depth. The resources of Bell Labs began to be shifted to Walter Brattain, a quiet experimentalist who'd been a rancher in Oregon, and John Bardeen, an even quieter theoretician from Wisconsin. (Bardeen was so quiet, and young-looking, that when he'd asked older students at the University of Wisconsin to play billiards for money they'd invariably taken up his offer. He was just as quietly polite when he took their money after the games—he was a superb player, and one of the best hustlers the campus had known.)

Now, in the fourth-floor offices of an innocuous research block in Murray Hill, New Jersey, the two friends Brattain and Bardeen used insights from Ohl and from chemistry teams at Purdue; they used translated documents recounting abortive German wartime research on semiconductors; they used their knowledge of quantum mechanics and of new chemical fabrication techniques, and they never let up. In October 1947 they were getting the first signs of success, and by December 1947 they were certain. They could make electrons start flowing, and they could also make them stop. They'd built the atom-level on/off switch Turing had sought.

It was one of the great discoveries of modern times. In all of human history, mankind's labor has been held back by the awful force of friction. Hoes scrape and drag along the ground. The slaves who built Egypt's pyramids spent almost all their energy in overcoming friction within their shoulder and leg muscles, or between the huge slabs they were moving and the ground underneath. Steam engines and car engines and even the fastest airplane's jet engines also waste tremendous energy on overcoming friction. But these silicon rocks can shift electric currents through in one direction or another, and the rock itself doesn't have to move, swiveling bodily like a big metal switch to one side or the other. That would be too slow and cumbersome. The rock can simply sit there, Buddha-like, and transform internally, allowing streams of electrons to be shunted down veins of modified ore inside it as needed.

If Turing wanted to send an electric current through, but only when a particular decision was made, he would just have to lead that current up to one of these special ore veins. At first the current would wait there, dripping electrons uselessly, unable to cross. But let that ore vein be transformed by the subtle techniques Brattain and Bardeen had developed, and then everything changes. The ore vein "tunnel" changes, and the signal can now roar along.

A new technology was born, which meant that someone had to come up with a new name. The general label wasn't too hard. When you're working to control large mechanical objects, it's natural to say you're working on "mechanics." Here, controlling individual electrons, the field had already received the understandable name of "electronics." But what should the key device that this technology used be called? This was a worrying moment, for engineers are often a menace when it comes to words. There were discussions, and votes, and one proposal was the not-quite-graceful "surface-state amplifier," while another proposal was the even harder to pronounce "iotatron" (from the Greek letter *iota*, signifying small).

Mercifully, though, John Pierce was brought in, one of the

lab's engineers, who dabbled in science-fiction writing and had a gift with words. He focused on the central insight. When the ore veins inside the silicon were "on," then lots of electric current could cross. When they were "off," there was high resistance to any current crossing. This meant that the device *transferred* a *resistance*, whence his more euphonic proposal:

"We have called it the Transistor, T–R–A–N–S–I–S–T–O–R," Bell's research director explained, at the press conference launching it on Wednesday, June 30, 1948. "It is composed entirely of cold, solid substances."

The first applications came easily. The Bell companies had long had a tradition of making devices to help the deaf—a legacy from Alec's love for Mabel—and it was natural to use transistors to make them smaller. (This was helped by the fact that John Bardeen himself was married to a woman who was hard of hearing.) Hearing aids are a bit like phones that have to be carried around. But ordinary telephones were clunky machines, dating from the Victorian era, with big wires and switches. They needed many trillions of bouncing electrons to transmit even a dim whisper. With transistors inside, however, it was possible to rely on far smaller batteries, for fewer electrons needed to be shunted around.

Computer designers learned what was going on, including Turing in England, as with his 1948 letter from Jack Good. In America the ebullient Grace Murray Hopper, working at Harvard, took such confidence from knowing that improved microscopic switches were coming up that it seems to have helped her in developing the world's first "compiler"—an indispensable part of modern computers, which translates the programmer's instructions into the arcane listing of switch positions inside the computer. She had followed women's basketball for years, and often noted how a forward pass took place: the person throwing the ball imagined the spot where she expected a teammate to catch it, and began the throw even before her target was there. Hopper liked explaining in later years how she'd used that image in the logic of her earliest compilers, to send out instructions that would be waiting in advance of the actual switch-shifting the computer was undertaking.

But when Hopper did her pathbreaking work on compilers in 1952—four years after the Bell press conference—she still had no workable transistors to use. The team that had been so successful in 1947 had broken up. Partly it was because Bell executives wanted solid, reliable components for their continent-wide phone switching system. Their goal was to have parts that would fail only once every twenty years, yet the first transistors failed almost every day. (An entire batch was once ruined when technicians touched it after opening a door: their hands transferred enough copper atoms from the doorknob to destroy the silicon's ideal mix of extra atoms which could slow or energize electrons as needed.) Bardeen and Brattain were discouraged by the lack of support.

They also suffered from being under the ostensible supervision of William Shockley at Bell, and Shockley was a man with very strange views. When he'd first met Bardeen's wife, Jane, he'd told her that his own children were inferior to him. Jane had demurred, thinking that her hearing aid had failed. But no,

Shockley explained, it was a fact, and the reason was that his own wife was genetically inferior to him as well.

When Bardeen and Brattain built their working transistor, Shockley was beside himself. How could these men have gotten there first! Bardeen wasn't the one who was supposed to come up with great ideas. The fact that Brattain, from the cattle ranch in Oregon—and so a cowboy! a hick!—had been part of the discovery made it even worse. Shockley tried to take credit for the work. At the 1947 press conference he hogged the microphone; when an electronics magazine came to take a photo of the great transistor discoverers, he pushed Bardeen and Brattain aside and sat himself at the lab desk they had used. Although he improved their initial ideas considerably, that wasn't enough; he wanted everyone to believe he had, basically, done it all. Bardeen left, and then others, and with no one remaining to yell at, soon Shockley left as well.

Although this meant no transistors were mass-produced in time to save Turing's life, it did have a remarkable benefit. For Shockley was such a good liar that when he left Bell Labs to create his own fortune in the apricot groves and scattered factories of a valley south of San Francisco, many of America's most skillful engineers and physicists wanted to work with him. He was, after all, the man on the cover of *Electronics* magazine, peering into a microscope at his desk at Bell Labs; he had been fêted at the press conference announcing the transistor; there were rumors that a Nobel Prize might soon be his. He also regularly pointed out how stupid the staff were at every competing company. What young engineer wouldn't want to travel to this semi-rural valley and become rich?

They came, they saw Shockley in action, and they fled. But while the engineers he'd forced out of Bell Labs dispersed across the entire country, those he forced from his new company (modestly named after himself) liked the California sun so much that they didn't go far. Shockley became a vast centrifuge, an inadvertent innovation machine. The bright people his reputation attracted quickly bonded with one another when they realized how awful he was, and kept those bonds when they were flung out to create their own firms nearby.

He managed to lose not just Robert Noyce, who co-created the modern technique for printing vast numbers of transistors on individual chips, but also Gordon Moore, who co-founded Intel, the most successful of all companies for fabricating those chips. Noyce became a millionaire, Moore probably a billionaire, while Shockley—never making any money, unable to get the ingrates who flocked to him to recognize his genius—kept on repelling more ambitious bright engineers, who joined forces with his competitors. The apricot groves had a new name: Silicon Valley was born.

And the world changed again.

New technologies can transform a society, and from that valley came new technologies galore. Without transistors and the fast-switching computers they made possible, we'd have no cell phones or weather satellites or spy satellites; no CAT scans or MRI or GPS; no cruise missiles or smart bombs; no solar cells or digital cameras or night-vision goggles; no laptops; no spam, but also no e-mail and no Internet; no widespread credit cards or cash dispensers or scanners or spreadsheets. There'd have been no mapping of the human genome; no plasma-screen TVs; no CDs or LEDs or DVDs or iPods or in-flight screen entertainment. There'd be no billionaires named Gates or Jobs; no—and how this will date us—Amazon.com or eBay or Google or Pixar.

At first it was easy to keep the world the way it had been, and just add on a few of the new items. But small gadgets can have unexpected effects. The first transistor radios went on sale in the 1950s, and the quantum effects they harnessed used so little power that their batteries were small. This meant kids could carry them around, which meant they no longer had to listen to the same music their parents did. Teenagers more and more formed their own subculture, and a new market for popular music was born. With cheap electric guitars and low-cost amplifying speakers—also made possible by silicon—small groups could match the volume of big bands. Obscure start-ups could flourish. Elvis and then Motown and the Rolling Stones appeared.

Transistor technology on its own didn't create rock and roll. There were lots of other trends at work—a big surge of young people coming of age in the postwar baby boom; a strong feeling after World War II that racism was unacceptable (which led to civil rights, and the sharing of black and white musical styles in Elvis's first Memphis studio); ever more suburbs and affordable cars for kids to party in. But electronics accelerated all those trends, melding them in a way that might never have happened otherwise.

The landscape changed. Huge chains of retailers could use computerized inventory control to fine-tune their offerings and to lower their costs in a way traditional stores couldn't. Behemoths such as Wal-Mart began clomping across the landscape; malls became virtually indistinguishable from one another as they filled with the same trendy chain stores.

Jobs changed. Executives found themselves checking their own spelling. Traditional manual occupations were sucked inside computer chips, and this changed the neighborhoods where those workers had lived. A man who works on the docks has clear rules at work, and passes on the notion of clear imperatives to his children. The children in that neighborhood might fight at school, but those fights also tend to have shared rules; teenage girls will be harassed, often roughly, but it's still often limited by shared principles, about dating, dress codes, and the like. Yet when the dock jobs go, replaced by computerized instructions for automatic cranes, children lose the role-model of parents who they see having to follow such rules. Solid, blue-collar jobs have crashed since the 1970s. Neighborhoods left behind have fallen apart. Fresh mixes have been created.

Even in richer areas, traditional notions of community began to fade. Radio and TV, when they began, sent out signals that flew equally in all directions, hence the word *broad*casting. This encouraged simple national brands and simple large clusters of consumers. Even when mail-order catalogs were delivered, there were only a few sorts, sent in huge batches that couldn't target niche customers. Computer switches, however, can quickly sort through many, many choices. This led to targeted direct mail (by the early 1960s), and soon after to increasingly special-ized radio stations, cable television stations, and the like. People no longer had to respond as part of a group. There was more nomadism, and more starkly individual choices of where to live, whom to marry, how to worship, and when to vote. Strange things happened. An older generation, which accepted that ex-ercise was something a few professional athletes might have to do, produced offspring who would go to large rooms, talk to nobody, and struggle to enhance their own utterly individual muscle tone.

Democracy changed. Before the computerized satellite links of the early 1960s, ordinary people never expected to see vivid, real-time footage of foreign disasters, rebellions, or famines. (All they did occasionally see were brief extracts, edited in occa-sional newsreels at the movies.) It was natural to defer to government leaders, who had their own superior sources of information—generally ambassadors or other emissaries, who communicated with them at relatively great expense by telex, telegram, or plane. But now? At the moment when a fresh tele-vision image rushes in from abroad, no one knows more than anyone else. A new mistrust of government was born—helped as always by other factors—and has persisted ever since.

Turing's progeny spurred one another on. Probably the last computers that could be entirely comprehended by one person were built in the late 1950s. But if you use a slide rule and a draftsman's drawing tools to work out the connections needed in wiring up a thousand switches for a computer, you're not going to go back to the slide rule when you want to try wiring up a million switches for a more advanced computer. You'll simply use the first computer to do it for you. Computers have begotten computers of escalating internal power ever since.

As a result, we now can detect such small numbers of electrons that we are able to see and hear things that in the past would have been utterly imperceptible. And we can use our control over those electrons—our understanding of how to make them teleport forward or seemingly halt—to bring their immense powers of speed and agility into our world. The door behind which the ancient electrical force existed has been opened one notch wider. This is Turing's legacy.

Consider the way GPS navigation works. Hundreds of miles above our heads, electrons are being led back and forth within the transmitters of GPS satellites. The shimmering force field that stretches from those electrons starts to wobble each time the electrons move, and that faint wobble continues, free-floating, all the way down to us on Earth. It's a GPS location signal, being sent on its way.

The wave that reaches the ground is invisible and inaudible to us: no one could peer at the orbiting satellites and see this wave coming; no one could cock his head and hear it.

Our eardrum is composed of atoms that have an immense number of electrons around them. The invisible rolling wave from a GPS satellite would be lost in all their jostling and shuddering. Even if the wave hit an old-fashioned TV antenna, it would have little effect; those metal rods, as placed on 1950s rooftops, still need a lot of electrons—perhaps several trillion—to be walloped forward in unison if there's going to be a noticeable response. That's better than our ears can do, but still involves more electrons than a satellite wave can produce.

Yet when that faint wobbling from space reaches our GPS receiver, something different happens—something made possible by the Bell engineers who showed how to access and control the quantum world. The incoming wave, feeble as it is, nudges forward a very small number of electrons, but those few electrons are guided to one of the special ore veins. The silicon they touch transforms; it's no longer a puritanical, disapproving, don't-even-think-of-letting-anything-get-across-me substance. Instead, its own electrons spring forth, stimulated by the tiny flicker of incoming electric current. Its inner cavities are no longer gloomy and blocked, but switch on and easily transmit a clear signal. The distant satellite has been heard.

Almost instantly the incoming ripple from space dies down; the ore vein inside the receiver shuts off. But then, a billionth of a second later, another ripple that's been floating down from the satellite arrives. The ore vein sparks into action again. After a mere few hundred billion repetitions or so, the distinctive identifying signal for that satellite is received.

A wavering bipedal human is likely to be scanning this GPS

device to "hear" a satellite hundreds of miles up in space, and so navigate her way to a new office building. She can then go on to use other devices for a Web search to examine, ever so easily, a few billion information sources scattered in silicon and metal memories within millions of computers worldwide. This too would have been inconceivable in earlier times, for even though there was a system of storing great amounts of information—the world's great libraries—it took trained archivists months to go through even one small part of those records.

That was because traditional information was stored by means of ink marks that have soaked into the thin slabs of modified wood pulp we call paper. But ink is big. Atoms are small. The archivists who looked through stored books, or even just the microfilm indexes for those books, were clumsily looking at immense solid lakes of electrons and other submicroscopic particles—for that is what written or typed letters, stretching for their great height of a quarter-inch or more, actually were.

Web searches are faster. Hit a key on a laptop or handheld browser, and long-stretching tunnels of electrons, in wires underneath our keyboards, are ready to begin the hunt. Our fast finger-pecks are ungainly, ponderously slow thumps to them, so there's plenty of time. (Every key on an ordinary laptop is sampled dozens of times each second, with electrons constantly reporting "No hit, No hit" to the computer's central processor, until, wondrously, one of the keys does slowly lower itself as we type.)

Great numbers of transistors in our computer and connected ones quickly reconfigure themselves to allow the desired request for Web information to pass, their ore veins instantly transforming into efficient conduits, then back into inert barriers, once the request has gone through.

Millions of Web pages begin to be scanned—some as summary indexes coded within a search engine's central computers, others at far-flung machines located all across the globe. Soon the request is making force fields shift the ancient yet powerfully charged electrons back and forth in thousands and then millions of pages, as the desired phrases being hunted for are compared with what's in storage. Decisions are made, and funneled through more transistor rocks, till finally the results start pouring back into our computers. The emissaries are reporting and a final set of signals is led to our screens. From their bizarre quantum flights, the answer we sought now glows into life.

It can all take place in the time we reach a hand toward a waiting cup of coffee. For centuries humans lived separated from the flurrying, fast inner world of electrons. But we no longer do.

The story of electricity still isn't over. We know that electrons can bounce along inside a wire; that insight led to telegraphs and telephones and lightbulbs and motors. A whirling, long-hidden force pushes those electrons along within the wires, and when shaken hard enough, the force can even vibrate as a wave that flies free from those wires. The result was radio, radar, and ultimately their miniaturized use in our cell phones. Quantum theorists found that electrons could teleport in great, precarious leaps, and even be forced to remain in seemingly immovable states of low energy; the result was switches contained subvis-ibly within solid rock, whence the computers transforming our lives. But beyond our technology, there's one event even more central to our lives where these electric charges so many billions of years old are crucially active.

PART V

THE BRAIN AND BEYOND

*T*he metal atoms that exploded from distant stars were not alone when they landed on Earth, back in the era billions of years ago when our solar system was new. Floating along with them in deep space had been carbon and oxygen and many other elements.

When those elements landed on the still-boiling Earth, only some of them fell deep beneath the surface. Many of the rest remained closer to the top, becoming part of landmasses and oceans, seafloors and swamps. Although they weren't metals, an electric field was leaking out of them as well, produced from the charged electrons and nuclei of which they were made.

Most of these nonmetals remained inert, stuck dully in their mountains and clays, but a few of them began to do something peculiar. The stretching electric field from their electrons made them writhe and twist and pull into strange configurations.

The sun over the new planet was hot. Energy was absorbed. The contorted clusters of atoms twisted some more, causing others around them to contort as well. Most of the shapes fell apart, but a few of the configurations created others so similar that they had duplicated themselves.

Life—built out of electric charges—had begun.

11

Wet Electricity

PLYMOUTH, ENGLAND, 1947

Electric forces weren't just important in assembling the first life, but are everywhere on our planet and in the body even today, active in the most mundane of activities. Just switching on a television or a computer, for example, means that from each glowing pixel on the screen, fresh electromagnetic waves are sent rippling outward at 670 million miles per hour. An awesome sequence of events quickly unfolds.

The watcher's eyes are likely to swivel forward in a sequence of stately turns as the screen's pixel glows: each quarter-ounce mass of eyeball tugged by six flat muscles, in a glissando slide within the slippery fat lining the orbital cavity. The eye blinks, the widened pupils are in position, and the incoming electromagnetic waves roar in.

Ripping through the thin layer of the cornea, they decelerate slightly, with their outermost edges forming a nearly flat plane as they travel inward, carrying the as-yet-undetected signal from the screen deep into the waiting human.

The waves continue through the liquid of the aqueous humor and on to the gaping hole of the pupil. The human may have squinted to avoid the glare, but human reflexes work at the rate of slow thousandths of a second and are no match for these racing intruders. The pupil is crossed without obstruction.

The stiff lens just below focuses the incoming waves even more, sending them into the inland sea of the jellylike vitreous humor deeper down in the eye. A very few of the incoming electric waves explode against the organic molecules in their way, but most simply whirl through those soft biological barriers and continue straight down, piercing the innermost wrapping of the eyeball, till they reach the end-point of their journey: the fragile, stalklike projection from the living brain known as the retina. And deep inside there, in the dark, barely slowed from their original 670 million mph, the waves splatter into the ancient, moist blood vessels and cell membranes, and something unexpected happens.

An electric current switches on.

Its existence seems odd, for the inside of the body is sloshing wet. We've seen electricity in telegraphs, telephones, lightbulbs, electric motors, radios, radars, and computers of every sort. But here too? Water and electricity are not supposed to mix. James Bond, famously, could terminate villains by tossing radios (electric) into their bathtubs (wet). Yet these tiny circuits in our eye sockets duplicate the operation of the most refined electrical receivers, despite being composed not of insulated copper wire, or even of cleverly modified silicon, but just of ordinary proteins, fatty cholesterol—and lots and lots of water.

Our entire body operates by electricity. Gnarled living electrical cables extend into the depths of our brains; intense electric and magnetic force fields stretch into our cells, flinging food or neurotransmitters across microscopic barrier membranes; even our DNA is controlled by potent electrical forces.

The result is that researchers today have created yet another new form of technology, a *liquid* technology, in which miniature puddles can be loaded with electrical particles that swarm into the inner recesses of our body. Anesthetics float down to electrical pumps in our nerve cells, numbing us so that surgery is endurable; Prozac latches on to electrical receiving units in our brain, keeping our sorrows in check; the electrically charged molecules that emerge from a Viagra pill operate on nerve firings elsewhere, making our pleasures rise. It's all part of the great shift in the frontiers of current science—from physics to biology; from the physical world outside, to the body and thoughts within.

This is so unexpected—what are live electrical circuits doing embedded in our bodies and brains?—that electricity's role in the body wasn't even imagined till recent times. Early Greek and Islamic investigators noticed a few electric effects, as we saw, such as the way the fur in pelts stands on end when you rub it, so long as the air is dry. Anatomists of the Renaissance and later found hollow white tubes running through the body and recognized that these were nerves. But they too assumed the nerves were run by a divine power or, in lieu of that, by miniature pulleys or perhaps hydraulic fluids, not by electric sparks.

What began to change this was the inability of most scientists, however independent they imagined themselves, to ignore the fashions of the society around them. Pumps were an exciting, rapidly improving technology in 1600s England and Italy, so when William Harvey was investigating the circulation of blood, it was natural for him to think of the heart as resembling a pump. Newton's followers naturally thought of the universe as a clockwork, because accurate clocks were a compelling fresh technology in the late seventeenth century.

In the early 1800s, many people had seen demonstrations of simple batteries and wires. The twenty-year-old Mary Shelley, sharing ghost stories one stormy night with friends near Geneva, had naturally thought of Dr. Frankenstein as using electricity to give his monster life. In the 1840s, telegraph lines were being built, carrying messages using fast-surging electricity. As the main western European cities were linked one after another, it was almost impossible to believe that the long nerves that carried messages inside our bodies didn't somehow surge with electricity as well. When German researchers managed to measure it clearly in the 1850s, they found that electricity didn't travel through living nerve cells at the millions of miles per hour that it traveled in telegraph wires. Instead something seemed different within the body, for the speed was only 100 miles per hour—just a few times faster than an arm moves in a fast jab.

In one sense that was satisfying, for it would be hard to understand how fragile human tissue could hold together if there really were signals rushing inside us at a million miles an hour. But it also was confusing, for the chemistry of that era still couldn't explain it. Eyelids are tugged down by muscles. Anatomists can easily find those muscles. But there are no little muscles acting as booster units to keep our numerous nerve signals going. And if our nerves were like telegraphs, where were the batteries and what, exactly, was inside the cables? There are no long strands of copper wire or any other metal inside our bodies.

The solution the researchers were looking for—the explanation of how electricity can exist even when surrounded by water—appeared only when they stopped focusing on the machinery of the era. Telegraphs work by bouncing electrons around, but electrons, of course, are just one part of the atoms from which they come. Telegraphs and lightbulbs and even computers have to rely on those small, vulnerable electrons. But our electrical technologies are a mere two centuries old. Evolution on Earth has been operating for billions of years, and long ago yielded another approach for conducting electricity, using not just tiny electrons but entire atoms.

The trick is to find atoms that have a greater-than-usual amount of electrical force leaking out. Normally we're taught that doesn't happen, for there's as much negative charge in electrons orbiting an atom as there is positive charge in the nucleus at the atom's center. The result is that the whole atom is in a state of equipoise. That's why it's electrically neutral, and why even the great Newton sometimes thought of atoms as boring, simple spheres.

Yet in quite a few atoms, such as those of the common metal sodium, it's easy to tug off a single outermost electron. Our planet and bodies are loaded with these amputees. That's most convenient, for this galumphy giant is ideal at pushing other electrical charges along. It has one more positive charge in its own center than it has negative charges in its remaining electrons, and so it beams out a strong positive electrical force field. Also, this big, atom–sized sodium chunk—lacking one of the orbiting electrons that ordinary sodium has—can survive in places where tiny electrons cannot. It's impervious to churning water or reactive oxygen; atom–sized ions like it can spend millions of years loose in the atmosphere, pummeled by wind and rain and electrical storms, or buried deep inside mountains, crushed under miles of rock.

Individual electrons couldn't survive for long in the warm, sloshing water of a living body, but these giant reshaped fragments do fine. Any atom that has a different number of electrons than it normally starts with is called an *ion*, from the Greek for "traveler." The snipped–down sodium atom is called a sodium ion.

This is what our bodies use to carry the currents Helmholtz measured. But how? Nerves are smaller than early anatomists thought they were; the white hollow tubes that Renaissance dissectors found are actually just conduits for the real nerves, which are much finer, almost like hollow threads, miniaturized far below ordinary vision. The narrowest part of all is the *axon*, the long-stretching part of a nerve cell along which signals shoot. They're so small that even with modern microscopes it's hard to see clearly inside most axons.

Luckily for science, different nerves send signals at different speeds. If a nerve is very slender, the signal goes fairly slowly. If the nerve is wider, the signal goes faster. This meant that twentieth-century physiologists who wanted to improve on what the first German researchers had begun simply had to find creatures that needed ultrafast nerve signals for their attacks and escapes, since that meant the creatures would likely have wide, plump nerves. The physiologists would also need pretty long creatures, since lengthy nerves would be easier to pull out. The logic's great, until one thinks what it means: hunting for huge, fast, living animals. Frogs would seem too small, bears could be too slow, but giant squid or, in lieu of that, ordinary squid—which need fast signals for their jet-propelled strikes—are ideal.

First, of course, one had to find one's squid. Alan Hodgkin, a gentle young English Quaker, had some difficulties when he returned to Plymouth, England, in the summer of 1939, after a stint in the United States. He went out on trawlers, he scoured the fish markets, but where were the squid? In chatty letters to his mother, he tried to be positive, but then, despondent, he couldn't help bringing up "the almost complete lack of squid." Yet by the end of July his luck had turned. He went away for a week's vacation to Scotland, having asked local fisherman to keep on hunting, and they came through: "On my return, found a large supply of squid waiting for me."

The squid nerves that he and his even younger colleague Andrew Huxley pulled out dwarfed anything from more-common creatures. These nerves were so big—almost as wide as a crisp pencil line—that they could stick a thin glass needle right down the middle of each one. (The squid was dead, but the nerves were "alive," in the sense that for several hours they still worked even without a host.) The nineteenth-century researchers could only measure along the length of a nerve, not see what was happening within. Hodgkin and Huxley, though, could now measure the electricity *inside* the nerve and compare it with the outside.

At first their experiments failed because the hollow needle scraped against the membrane. But Huxley was good with his hands, and eventually, with the help of some miniature mirrors to see upcoming bends, they could steer the needle without scratching the fragile, still-living nerve.

In the first few weeks, using a time-honored sophisticated technique of neurophysiologists, they squished the axons to squeeze out the axoplasmic goop inside. They didn't find many of the atom-huge sodium ions there, which was intriguing, since there are plenty of sodium ions in seawater and in blood; sodium, after all, is just part of ordinary salt (sodium chloride). The salty taste of seawater or blood, be it squid blood or human blood, is a sign of those ions in operation. Something inside the membrane of the squid's axon was taking sodium ions, those huge modified atoms, and grabbing on to them, and pushing them right through the membrane, so that they built up on the outside of the nerve.

This was great; this was seeing details that the German researchers of the 1860s had only guessed at. The squid was stockpiling sodium ions outside the membrane of its nerve cells. But, why? The two young men had a hunch—they'd studied physiology at Cambridge with the world's experts—but before they could go any further, the war came. Hodgkin worked on radar, Huxley with the Admiralty, and it wasn't until 1947 that they could get back to full-time research. Hodgkin's new wife, Marni, wrote to her parents:

"Alan . . . is like a dolphin that has suddenly been released. . . . He plunges and gambols and cavorts in pure research after so long. . . ."

It was frustrating to have stopped, but the years in radar hadn't been wasted. Hodgkin and his wartime colleagues had repeatedly used the old insight that it's easy for electric currents to travel along a smooth, wide path. There are usually plenty of available electrons in such a big path, and so there's not much "resistance" in the way. Narrow paths, however, are harder for electric currents to continue along. The electric current faces more resistance. Since most nerves are very narrow (notwithstanding the relatively enormous nerve of a squid), they offer immense resistance to any electricity trying to squeeze through. This means, as Hodgkin explained a little later:

"If an electrical engineer were to look at the nervous system he would see that signalling electrical information along [narrow] nerve fibres is a formidable problem. . . . The [nerve] fibre is so small that . . . the electrical resistance of a metre's length of small nerve fiber is about the same as that of 10,000,000,000 miles of [thicker] copper wire, the distance being roughly ten times that between the earth and the planet Saturn.

"An electrical engineer would find himself in great difficulties if he were asked to wire up the solar system using ordinary cables."

Nerves have to work differently. They can't pour their electricity straight down the middle of their axons, as Alexander Bell had imagined electric sparks rolling down a copper wire in his telephone. Instead, nerves have somehow to be powered up from the side, getting regular boosts from something very strong and stable waiting here. It would be as if an engineer knew he or she was going to have problems keeping a signal going in a long wire, and so thoughtfully installed, oh, a few trillion booster units at regular intervals along the edge.

That's what the huge sodium ions were doing. They were there to keep a nerve signal going. (Further research showed that potassium ions are also central to nerve conduction, but since their operation is similar, the text will stick to sodium for clarity.) When we have a thought, and a nerve cell in our brain begins to fire, the signal will die out in a fraction of a millimeter unless some of this charged-up sodium squeezes back in from outside to keep it powered along. What Hodgkin and Huxley made clear—and what earned them a Nobel Prize—was that the cell membrane isn't a continuous rubbery barrier, impermeable and secretive, keeping our thoughts locked in coy Freudian depths. Rather it has lots of little gaps that widen to let the sodium ions go through. Not a lot has to pour across, just a few thousand sodium ions along any millimeter of the nerve, but that's enough.

Anything that lets the sodium boosters in will start a signal. In an eye peering at a computer monitor, for example, the incoming electric waves from the screen hit ingeniously shaped molecules known as rhodopsin, which exist on the retina. Imagine the rhodopsin molecules as palm trees. The molecules twist like palm fronds in a typhoon when the light hits, and a part of the rhodopsin—its "roots"—pull upward. Since the rhodopsin trees stand in a slurry of sodium ions, gaps open at the base of each one as the trees lift up. The sodium pours through those new gaps into the nerve beneath, and the signal begins.

That first jolt of sodium, entering the very start of the nerve, makes the *next* millimeter of nerve membrane do something strange. It warps and bubbles and contorts and then, suddenly, holes start opening up there, letting more sodium that's been waiting farther along on the outside pour in. When that booster sodium arrives, the next section of the nerve has the strength to bubble with opening holes. Sodium that's been stored on the outside even farther along pours in, and the sequence repeats, rippling quickly along the full length of the nerve.

After the whole signal has traveled along, the nerve is a

bedraggled mess, pierced with holes and sloshing with sodium. Before it can fire again, it has to rebuild itself, which means bailing out the extra sodium that poured inside, and closing up the holes. That's so energy-consuming, as is keeping the electrified sodium on the outside, and not letting it dribble in while waiting for another signal, that 80 percent of the energy that goes to our brain—all the sugar and oxygen; all the nutritious residue of steaks and muesli and Frosted Flakes and Junior Mints—is devoted simply to fixing this sodium-hole damage.

Sometimes the nerve doesn't recover as quickly as usual. When the air is cold, your fingers get clumsy. That's because the fatty nerve coverings in your fingers start to thicken, much as greasy lamb's fat congeals after a meal. As a result, the sodium pumps within the nerves that go to your fingertips don't work as well as they do in warm weather. It's why we need to warm up before any task that requires fine motor control. Glenn Gould, the great pianist, was often restless before playing, until he could find a deep sink or bucket where he could plunge his arms in hot water. Critics sometimes made fun of him, but then they heard the results. Once his congealing fat had softened, once the path for his electric-powering sodium ions was ready, Bach's masterpieces were his. (Ice cubes help keep down the yelps in earlobe-piercing for the same reason.)

Sometimes the problem is more serious than a gust of chilled air. The liquid called tetrodotoxin is one of the gravest nerve poisons that exist. When it splashes around a nerve, it acts on the sodium pumps, closing them tight. Were this to affect only a few nerves, such as the ones involved with our looking around, befuddled, to find the exact location of the TV remote control, that might not be too bad. But there's a great similarity in nerves throughout the body. As the tetrodotoxin seeps everywhere, the nerve signals traveling to the heart and lungs are affected as well. We might have some awareness of this, we might look down and quite sincerely want our nerves to keep firing, but with the sodium channels blocked and none of those booster ions falling in, the electrical signal would soon peter out. The result would be death by suffocation. Tetrodotoxin is produced in its natural state by the dreaded Japanese puffer fish and duplicated by eager chemical warfare specialists worldwide.

Alcohol is a bit in-between. It makes the fatty nerve membranes begin to congeal, but not enough for instant death. The result is more like cold-finger fumbling, only this time the membrane disruption operates on nerve cells deep inside the brain, carrying our thoughts and memories, rather than just in the extremities. For those who, as Samuel Johnson observed, wish to escape from themselves, it's consoling to temporarily weaken the walls that hold their electricity-boosting sodium pumps in place.

Technology regularly advances faster than science, and people happily used alcohol and—if wise—avoided puffer fish, for thousands of years before the details of sodium pumps were known. Early anesthetics were similar. They were greatly needed, for alcohol barely works in stopping surgical pain, and even into the mid-1800s the big teaching hospitals had to employ burly "grippers," such as ex-stevedores or boxers, whose job was to stride after escaping surgical patients and drag them back to the operating theater for their ordeal. (Flaubert, son of a surgeon from this pre-anesthetics era, gruesomely described what surgery then meant in the leg-amputation scene in *Madame Bovary*.)

The change with anesthetics came in the early and mid-1800s, when various helpful gases, such as ether, were found to knock out patients without too much fatality. Sigmund Freud, as a young medical student in the 1880s, was especially fond of experimenting with the properties of the modified plant extract called cocaine. It was excellent for eye operations, and also pleasing for the surgeons who sampled it, often repeatedly, just to confirm that the doses were right.

Only with the work of Hodgkin and Huxley was it clear, though, how anesthetics might work. Much as with alcohol, many of these molecules sink into the fatty membranes of our nerves, and diminish the working of ion pumps along nerve axons wherever they arrive. Pliers might be savagely tugging at a molar, or a needle might stitch together pieces of living tissue, but the nerve impulses with which the brain would recognize these iniquities will advance barely a fraction of an inch before— sodium boosters off—they dribble to a halt. As the details of how they work became known—and the different modes of action of general versus local anesthetics became clear—medicine improved tremendously. Serious operations such as heart bypasses became possible; keyhole surgery on knees became possible to control. Victorian engineers had only big electric engines and used them to move elevators or to run machine tools or refrigerator pumps. Today's biological engineers can use the microscopic electrical pumps of sodium ions to control far more delicate maneuvers within.

Hodgkin never got around to inebriating his squids, but colleagues did ply them with tetrodotoxin (which sounds unkind, but, given that the nerves being used had been removed from the squid's body, the objection is perhaps moot). When they poured tetrodotoxin directly on nerves to stop the sodium entirely, and then watched as it wore off, they could see exactly how the sodium pumps began to operate again. What they discovered was humbling to anyone who takes pride in our distance from tentacled, wide-eyed marine creatures. The mechanism was exactly the same as in humans.

"Considering," Hodgkin said, "that the squid is a very distant relative of man—our last common ancestor . . . died several hundred million years ago—this similarity in behaviour points to the survival value of the sodium channel in the animal kingdom." There is a sensible logic here, for living organisms—humanoid or cephalopod—have no choice in what materials to work with. Ions are an excellent, jury-rigged way of using electric current to send signals. What worked in the year 300 million B.C. holds up today.

12

Electric Moods

For all those hundreds of millions of years, electrically think-ing life-forms have wriggled, run, slept, procrastinated, or otherwise occupied themselves on our planet. Within each of these creatures, electric signals have been shooting along mem-braned nerve tunnels, like the world's most intricate roller coaster, its cars all lit up and racing along at night.

But in time each signal reaches the end of its nerve fiber. This presents a problem, for nerves don't form a giant tube network, plugging continuously from one into the other. Instead there's a gap between any two nerves. This gap, called a *synapse* (from the Greek word, *synaptein*, meaning "to fasten together"), was clearly seen as early as 1897. It's not a very large separation, just a few thousandths of an inch, but on the microscopic level it's a near-oceanic fissure.

How can the signal cross it? The solution would be the next great step in understanding our nerves and mind. Electrons would be swallowed up in the gap between nerves, and even single ions would be as useless as great beach balls bobbing in a sea. Yet microscopists knew that something had to make it across. They even suspected it was electric. But if it wasn't a small electron, and it wasn't a large atom ion, what was it?

The answer came from a forty-seven-year-old man, a pharmacologist at the University of Graz. Late on Easter Eve, 1921, he woke up in the night, suddenly realizing exactly how those gaps are traversed. This was fabulous; science's understanding of the nervous system would now be complete. He switched on the light, wrote down his great insight, then fell back asleep. In the morning he woke up again. He'd always been a promising researcher, but what he'd dreamed of that night: well, it was an idea for the ages. He looked at the piece of paper where he'd written his great idea.

And he couldn't read it. Penmanship had always been one of Otto Loewi's fortes, but not at three a.m. The next day was one of the worst of his life. However much Loewi stared, he couldn't make out his scrawled words on the slip of paper he'd used. Nor, however hard he tried, could he recall even a fragment of what he had dreamed.

The following night, Sunday, he delicately let himself fall back asleep. If he was lucky, the answer would come back. Midnight came, he was still asleep—no dream. One a.m.—no dream. Two a.m.—still no dream to wake him. But then, as Loewi lovingly remembered: "at 3 o'clock, the idea returned. It was the design of an experiment."

This time he wasn't going to trust his idea to pen and paper. Loewi got dressed and hurried along to his laboratory. He had a way to identify the substance emerging from a nerve! What Loewi realized he had to do—squeamish readers might wish to skip the next few paragraphs—was to kill two frogs and cut out their hearts. He'd keep one heart attached to the nerves that poured their unknown chemicals down on the heart. He'd watch to see how the heart behaved—whether it slowed down or speeded up—as he made more of the chemical come out of the nerves. Then he'd siphon that unknown chemical so it poured over the other heart. If the second heart reacted in the same way, then he'd know that something in that liquid carried the answer.

Loewi could do this, for, like many anatomists of his time, he had a ready supply of unfortunate frogs at hand, and he knew that even a deceased frog's heart would continue beating for quite a while. He got to work with his scalpel and soon had the two hearts in two separate buckets, where they throbbed away. He squeezed on the big vagus nerve leading to the first frog, so that more of whatever liquid it had inside came out. The first heart slowed down. He then led some of that liquid into the second bucket, where the second heart was still pulsating away, entirely on its own. A few moments after he poured in the liquid, the second heart also began to slow down. The liquid that came out of living nerves really was powerful enough to do that.

What Loewi and his successors realized was that inside these liquids that squirt between our nerve cells there are relatively ponderous floating molecules. They're often built of several hundred atoms stuck together, making them much bigger than the bobbing sodium ions, so they survive their journey unscathed. They act like miniature submarines. Whole navies of them are released from tiny bubbles in the tip of a firing nerve cell, and they float across the synapse toward their target. Similar molecules exist at nerve junctions throughout the body, and are especially important in the brain's nerve connections. Since the brain cells we think with are called neurons, these molecules that transmit signals between them are called neurotransmitters. God stretches a forefinger to Adam on the Sistine ceiling, and His nerve endings dribble molecules across which—through the ingenious expedient of opening Adam's sodium channels—make the first man's nerves start quivering with electrical surges.

This is how signals cross the gap. Electric signals move along in a nerve to the gap at the end and make a powerful liquid come out of the nerve tip, and that liquid crosses the gap, and enters the next nerve, and carries along the message that the first nerve had sent.

If every one of our nerve cells sprayed out the slow-down liquid Loewi found, however, we'd be in trouble. We'd think a thought, or try to move our arms, and everything would start going slower and sloower and slooooooower. Luckily there are other types of transmitter liquids in our body. Some speed up the cells they reach, while others simply help them form new connections—dozens have been found by now. (One of these additional transmitters that Loewi also helped comprehend was initially called *Acceleransstuff* for its action in speeding up the cells it reached. We call it adrenaline.)

Each of the neurotransmitter submarines has a different shape, and if it finds the right berth for itself, it gets tugged right in. That tug comes from the same sort of static electricity that gives us a shock on a dry day. Several regions on the transmitter molecule have extra negative electric force (from a concentration of electrons there), while matching regions on the target nerve cell have extra positive electric force (from a relative lack of electrons). When the correct two regions get close, it's as if deckhands were pulling on docking ropes. They snap together tight.

If the process ended there, we'd be in trouble once again. For when the neurotransmitter arrives, the nerve it hits can now start firing, with its sodium pumps opening wide. But if the transmitter remained stuck there headfirst, that firing wouldn't stop. The signal from the past would keep on recurring. You'd have no way ever to get new sensations from the outside world, nor would you be able to create a new thought. You'd be stuck in that single moment forever.

Luckily there are yet other molecules operating in the gap between nerve cells in our brain and throughout, and their job is to act as demolition crews, disassembling the neurotransmitter almost as soon as it's made the journey across. In a most convenient burst of ecological efficiency, when they've torn apart the transmitter to its constituent parts, they then generally lead those parts back to the original sending nerve, where the parts are reabsorbed, rebuilt, and—all memory of the previous voyage wiped clean—led toward the surface to be ready to be sent out again. Electric forces pull all those parts along. Without that power, none of this would happen.

Lots of mysteries became clear after Loewi's work. Caffeine

has been drunk for centuries, with commentators even in the 1600s complaining that young scholars used it to excess, to stay alert when they were trying to catch up on their studies. But no one knew how it worked. Understanding electrical connections at the level of cell surfaces in the brain changed that. A common transmitter between brain cells is the bulky molecule called adenosine. When it reaches its target brain cells, it tends to slow down their rate of firing. What caffeine does is slip into adenosine's docking sites. With those berths now full, the adenosine can't get in. We might be exhausted, we might be craving deep rest, but with our receptor brain cells speckled with caffeine, our desperate adenosine outpourings can't find enough free berths to latch on and so can't get those cells to slow.

As the years went on after Loewi's discovery, even more precise details became clear. Nancy Ostrowski was a young American who'd once considered being a nun, and by the 1970s had switched to research. But she seems to have channeled some of the moral strictures from her previous life into her new work. In her laboratory outside Washington, D.C., she would get something like a small guillotine ready, then encourage mice to have sex, and then decapitate them while they were thus engaged. With sufficiently quick blenderizing of their brains, she found that many of their brain cells had been pouring out endorphins. These are naturally produced neurotransmitters shaped much like heroin or morphine. When they cross synaptic gaps and dock with waiting receptors, mammals experience great pleasure.

Endorphins are temporary, but our overall moods are more

long-lasting. Some people are like P. G. Wodehouse's Bertie Wooster, for whom cheerfulness kept on poking in. Others are the opposite, and quite happy to wring the necks of all those who constantly tell them to be more chirpy as they go through the day. The more precise our knowledge of how electrically charged molecules and ions move inside the brain, the more easily we can reach in and control our moods, and to some extent even our temperaments.

This is the latest stage in using the understanding of electricity for a new technology. The consequences of the telegraph and computer took decades to become clear; the future that insights into neurotransmitters will create still remains to be seen. The great breakthrough in using them for detailed psychological control came at the Indianapolis labs of the Eli Lilly company in the early 1970s. Many researchers had known that the neurotransmitter called serotonin is important in setting our moods. There are many qualifications, but roughly, individuals who have low amounts of serotonin in their brains often feel depressed. Yet how to control it? Blasting the brain with a powerful chemical such as Thorazine will make us a bit better, but Thorazine, alas, has the accuracy of a flamethrower and attacks many other useful circuits in the brain. The result, when Thorazine alone was used, were mental hospitals where patients could be taken out of their straitjackets, but only because they could be trusted to sit vacantly in lawn chairs, having lost much of their personalities in the process.

What the Eli Lilly researchers found was an ingeniously indirect way to raise the level of serotonin and nothing else. They couldn't make more serotonin come out, but they could make the little bit that did appear stay in operation much longer. For serotonin levels are controlled not just by how much is poured out from firing nerve cells in the brain, but also by how quickly the demolition crews—the janitorial molecules—floating in the gaps between brain cells grind them apart and lug them back to the transmitting cell to be reabsorbed. If someone isn't producing enough serotonin, or if their receptors for it aren't working well, why not just slow down the demolition and reabsorption process? Prozac pours out small, electrically glistening molecules that latch on to several of the molecules active in the demolition process, and stop them from working at full efficiency. The result? With the demolition molecules out of commission, the small amount of serotonin that's naturally dribbling out doesn't get destroyed so quickly. Its levels remain steady, or even rise. We feel better.

The Scottish philosopher David Hume used to wonder what it would be like to be the manager of a playhouse, standing backstage and watching all the different characters racing about. By those characters, he meant the different aspects of his personality. He lived around the time of the American Revolution, when Volta, who started off our story, was still a young man. So Hume had no idea that all those characters were held inside his mind as patterns of electricity shooting along the long cables of his neurons, or as electrically charged molecules floating across his synapses. But I think he would have liked the idea that in the two centuries since his life, we would create a civilization powered by electric forces, and we would learn, near the end of that period, that the machine that lets us see and understand it all—our brain—is electric at its core.

Once again, the extremely small size of the ancient electrons we're composed of is central to its working. Electrons are so small that the molecules they help create are still far below the size scale we can see directly. That means immense numbers can fit inside us. Our brain weighs only three pounds, but has about 100 billion active nerve cells nestled within it. This provides as many electrical signaling stations as there are stars in the Milky Way galaxy. Signals flash through our nerves at about 100 miles an hour; they take merely a few thousandths of a second to cross the synaptic gaps and continue in the next nerve. Since the sodium pumps and neurotransmitters move faster than rocks fall or trees snap in the world outside, that's why we can use their odd constructions to keep us intact as we dodge and stumble our way through the world.

Electricity's speed lets us survive, but if this were its only strength, we'd be utterly lost and without memory. For sensations, of course, are constantly pouring into us in quick succession. Great barrages of air molecules hit the taut skin of our eardrums, and when the barrage is especially intense, we experience it as a clear sound. Our nerves carry it deep into the brain, with the tiny sodium pumps boosting the signals, and bobbing neurotransmitters floating upward to carry it even further. Yet if the next barrage of sensations from outside took over those same nerve cables, the first signal would be automatically superseded, wiped out of existence by a new configuration of sodium-pump firings pouring in after it. We'd have no memory of what had just gone before.

The ancient stability of electric charge is what saves us. For remember the auralike force fields that emerge from all the charged particles on Earth. They have existed, waiting, for billions of years. They're very strong, and very old, and when whole systems of neurons are arranged to fire in particular patterns, those strong-pulling force fields will hold those patterns intact.

Short-term memory flickers might disappear in just seconds or minutes, but our deeper memories, the ones that constitute our very personality—the permanent actors in David Hume's drama—can exist, impermeable, sustained by the arrangement of electric whirling pulls within our cells, for hours and months and then decades on end. A young woman meets a wonderful man and is swept off her feet. Decades later, old and bent, with grandchildren around, she hears one of her children read aloud from his love letters. At first the words are distant, scarcely recognized. But then the sodium pumps and neurotransmitters with their glowing electricity stir into action. She looks up.

She's remembered.

*T*he universe is very old now, and the original electric charges from the Big Bang have long since dispersed. Many of the individual charges were destroyed as they traveled the galaxies, but in their place—always—new charges were created. No exception has ever been found; the sum of electric charge in the universe has never changed.

Dawn has come over raw, molten planets, and ingeniously electric-sealed living molecules have evolved and reproduced. Assemblages of self-conscious neural cells have taken shape to become sentient brains, and electrically operating retinal cells have guided mobile beings.

In all those lives, in all those eons, this one constant has remained. All the firestorms and cyanide apples and telegraph messages have existed simply because of the shifting from place to place of those charges. Sometimes those electric charges have been channeled down copper wires; sometimes they've moved through the neurons of lovers, or students, or wild-eyed political demons; at yet another time, in the future, electric and magnetic fields from our destroyed sun will be sent barreling across the galaxy in turn, carrying unheard messages to far distant stars. We are fragile organisms, living amid these roaring, stately, powerful migrations of electrical charge.

The dominion of electricity shapes us all.

"I shall conclude this chapter of precepts, with advising all young electricians to be exceedingly cautious. . . . A large shock . . . might affect his intellects in such a manner, as they should never be what they were before."

—JOSEPH PRIESTLEY, *Familiar Introduction to the Study of Electricity*, 1768

WHAT HAPPENED NEXT

JOSEPH HENRY became a friend of Abraham Lincoln and died in 1878, revered as America's greatest scientist. He rarely spoke about Morse, but did once remark: "If I could live my life again . . . I might have taken out more patents." SAMUEL MORSE nursed his vast fortune, but was dismayed when the American Civil War led to slavery's abolition. He died in 1872 of what seems to have been a stroke, increasingly frantic that Henry might be shown to have been the telegraph's inventor. In the mid–1990s, international authorities officially discontinued Morse Code for all military or maritime use.

ALEXANDER GRAHAM BELL retired to Canada, where he became a pioneer of research on flying vehicles and high–speed hydrofoils; he was an early proponent of women's rights. There's a photo of him as a very old man with a white beard, standing on a pier in Nova Scotia, watching a test run of his most advanced hydrofoil, a shiny, streamlined aluminum blur, heading toward a speed record. His wife, MABEL HUBBARD BELL, can't be seen in the picture because she's piloting the hydrofoil.

THOMAS EDISON continued as an inventor, but his youthful creativity left him only a few years after his success with the lightbulb. He lost a fortune in ore-extraction ventures, and then in trying to construct a concrete boat. At one point he controlled the key patents in cinematography, but then decreed that no film longer than twenty minutes could be made, resulting in the collapse of the then-flourishing New Jersey and New York movie industry, with most directors fleeing to distant California. When he died in 1931, President Hoover asked that all lights across America be dimmed at 10 P.M. on the day of Edison's funeral.

The man who discovered the electron, J. J. THOMSON, never drove a car or traveled by plane in his life, yet the Cavendish Laboratory he directed became the world's greatest center for experimental research, leading to fundamental discoveries in subatomic structure and much later to helping identify the structure of DNA. In old age he took to playing golf alone, so he could stroll over the links looking for wildflowers; he told his son that if he had his life to live over again, he would choose to be a botanist, for "so much potentiality in a tiny seed [is] . . . the most wonderful thing in the world."

The young boy who'd first scampered up the Royal Institution's steps as a teenager, MICHAEL FARADAY, was still regularly there more than half a century later. "Next Sabbath day (the 22nd) I

shall complete my 70th year. I can hardly think myself so old."
When he could no longer do research, he spent hours watching
the sky from his window, especially delighted when he could
see lightning.

With the great wealth he earned from the Atlantic cable, CYRUS
FIELD returned to New York for good and invested in the city's
elevated railroads. He was double-crossed by his business part-
ners, and then his son stole nearly all of what was left. He died
almost penniless. When he was very old, Field had a recurrent
dream that his ships had finished laying the Atlantic cable, yet
he had been left alone on the shore in Ireland. The ATLANTIC
CABLE of 1866 remains on the bottom of the ocean, long since
abandoned. Drifting ions in the Atlantic's depths cause occasional faint electrical currents to eddy into existence within.

The young swimmer and athlete WILLIAM THOMPSON ended his
life as the white-bearded, wise Lord Kelvin. As a strong believer
in religion, he noted that the sun couldn't have existed long
enough to allow the slow evolution that Darwin had postulated
to take place—unless some energy source beyond what he or
any of the other Victorians knew of was found to exist. Shortly
before his death, Becquerel's discovery of radioactivity and Marie
Curie's studies with uranium ores verified his modest guess.

HEINRICH HERTZ's last name was adopted for the international
term denoting radio frequency, and appears as the symbol "Hz"
on the world's radio dials. The little girl born to him in October

1887, JOHANNA HERTZ, had to flee Germany in the 1930s because her father had been half-Jewish. She and her younger sister spent years collecting and editing their father's journals and
letters for publication. GUGLIEMO MARCONI became a recluse,
living almost entirely on his oceangoing steam yacht *Elettra*. He
was a key financial supporter of Mussolini's Fascist party.

After successfully suing the postwar British government to get
increased payment for his contributions to radar, ROBERT WAT-
SON WATT left England in disgust and retired to Ontario. When
he was stopped by a traffic policeman for speeding one day in
the early 1950s, and it transpired that the police force used radar
guns to measure automobile speeds, the event became headline
news around the world, helped by Watson Watt's enthusiastic
explanations to visiting reporters. The town of SLOUGH, from
which he'd escaped in 1936, survived World War II sufficiently
intact to inspire successfully creative loathing in yet further generations of British citizens. When the writers of the BBC satire
The Office were looking for the optimally bland town to use as a
setting for their program, Slough was the natural choice.

HUGH DOWDING was forced out of the RAF after his success in
directing the Battle of Britain. In his old age, he believed that the
spirits of dead Battle of Britain pilots were communicating to
him through angels. COLONEL WOLFGANG MARTINI, the Nazi
signal officer who told the Luftwaffe it could disregard Britain's
radar installations, became a respected NATO officer after the
war. In the 1950s, at the Farnborough air show, he met Edward

Fennessy, one of the British engineers responsible for the Chain Home warning system. "I asked him why they didn't attack the radar stations," Fennessy recalled, "and [Martini] said Well they were not operational. [I said] so how did we track the Graf Zeppelin, and he nearly shot out of his chair, You tracked us?!"

CHARLES W. COX, inadvertent hero of the Bruneval raid, returned to Wisbech, in East Anglia, and opened a successful radio and television shop. The raid's success ensured the continued survival of Britain's then-experimental paratroop units, and the British Army's Parachute Regiment has the name of Bruneval inscribed as the first battle honor on its Colours. The main French Resistance volunteer, ROGER DUMONT, whose scouring of the site made the airborne raid possible, received an exultant message from Britain immediately afterward reporting its success. The message ended up being intercepted and decoded by German officials. Dumont was identified, tortured, and murdered. An hour before his execution, he wrote his family, "All that I have done I have done as a Frenchman. I regret nothing."

ARTHUR HARRIS, head of RAF Bomber Command, moved to South Africa and then returned to England, where he lived to a contented old age. Kind to his grandchildren, he was a great supporter of the Boy Scout movement.

The city of HAMBURG was rebuilt.

A few surviving WÜRZBURG RADAR sets were taken to Britain and used in astronomical research, where they helped create one of the first radar maps of the galaxy. The chaplain who'd protested Harris's policy at Bomber Command, JOHN COLLINS, became a critic of apartheid after the war, and went on to help found CND, the Campaign for Nuclear Disarmament.

ALAN TURING's mother wrote a privately printed biography of her son immediately after his death, but his memory soon faded from the history books, and the computer he had tried to build at Manchester was never a great commercial success. Only in the 1970s, as his Bletchley Park work began to be declassified, was he rediscovered by biographers. Today the most honored award for achievement in computer science is called the Turing Prize. Reflecting on his life, Turing once wrote: "As I have mentioned, the isolated man does not develop any intellectual power. It is necessary for him to be immersed in an environment of other men, whose techniques he absorbs during the first twenty years of his life. He may then perhaps do a little research of his own."

WALTER BRATTAIN left Bell Labs to teach at the small college in Oregon where he had first been a student in 1920. He remained modest about his great achievement, though he did once suggest that rock-and-roll musicians used his transistors for more amplification than he'd intended. JOHN BARDEEN moved to the University of Illinois and earned a second Nobel Prize (for work on superconductivity), the only physicist ever to be so honored. He remained even more modest than Brattain: one long-term golfing partner at the university asked him, after years of playing together, just what was it he actually did for a living.

After the failure of his Silicon Valley ventures, WILLIAM SHOCK-LEY abandoned all science research. Shunned by his professional colleagues for his increasingly racist views—and divorced by the wife he'd deemed inferior—he began donating sperm to a foundation designed to create genetically superior white children.

ALAN HODGKIN, the young Quaker who'd been distressed by his inability to collect squid neurons in the summer of 1939, became a leader in research into the cellular basis of vision, and crowned his career by becoming president of the Royal Society. His fellow young researcher ANDREW HUXLEY became a leader in biophysics, helping create the modern understanding of muscular contraction. At the time of this writing, he remains an active fellow of Trinity College, Cambridge. In 1938 the Jewish OTTO LOEWI was forced to leave Austria; he also was made to transfer all his Nobel Prize money to a Nazi bank. Welcomed in America, he became a U.S. citizen, and in his old age greatly enjoyed visiting the museums in his new home city of New York. He died in 1961, forty years after his great Easter Eve dream.

MR. AMP, MR. VOLT, AND MR. WATT

The world is made of electric charges, and our technologies operate through electric charges, and even our brains are powered by electric charges. Yet how to measure the flow of electricity? Three historical characters have had their names immortalized in the familiar units *amps*, *volts*, and *watts*, which describe—and summarize—what's happening inside all the electric devices we use.

The first was the French mathematics professor André-Marie Ampère, who in 1820 studied the way that rushing electrical currents created magnetic pulls. It was one of the only satisfying moments in his life—on his tombstone he had the phrase *Tandem felix* (Happy at last) inscribed—but in 1881, long after his death, the word *amp* was taken to measure the flow of charged particles. In ordinary household circuits, it's simply a count of how many electrons are passing through a given point in a wire each second. When 6 quintillion (6,000,000,000,000,000,000) electrons pass in a second, then we say that one amp of current is flowing; when 12 quintillion electrons travel through the wire, we call it two amps of current. An electric bulb lights up when a single amp's worth of electrons—6 quintillion individual electrons—are passing along each second. Switch on a car's ignition, and 50 amps' worth of electrons—300 quintillion electrons—pour into the spark plugs each second.

The second character was the prickly Italian Alessandro Volta. He created the batteries that allowed other researchers to investigate the force that can "push" all those electrons along, though, as we've seen, he had little understanding of why those batteries worked.

Michael Faraday's explorations helped make sense of that pushing force, and the key concept for measuring it came from Macquorn Rankine, a friend of both William Thomson and Faraday. Rankine had spent years designing railroads for the rolling hills of Scotland, and he used that experience to help come up with the notion of what he termed "potential energy." A train high on a hill might not be moving fast, but it has a lot of this potential energy, for once it starts down the hill it'll pick up speed thunderingly fast.

Rankine and Thomson used that vision to try to look into the force fields they imagined stretching out from batteries and other electric sources. In their mind's eye, there were invisible "hills" where the field was strong, and "valleys" where it was weak. Put a charged particle where the field is strong and it'll hurtle away, like happily startled train passengers who see their locomotive leading them fast down to a waiting glen. A "volt," Rankine and Thomson realized, should simply measure how steep—and so how strong—this pushing force is in the topography of one of Faraday's invisible fields. An amp measures how densely a current of electrons is flowing; a volt is the "downhill" pushing force that creates this flow.

The last of the three characters was James Watt, yet another ingenious Scot, and the patron saint of electric bills. Although he hadn't quite invented the steam engine, he had improved it greatly, and decided that his major achievement would be to persuade tight-fisted mine owners to actually buy this new machine.

What he needed, he realized, was a risk-free offer, showing that their savings from the new machine would more than offset the cost of buying it. But for that he needed to measure the savings, which meant he had to find a catchy way of summarizing what the horses dragging the carts and driving the pumps in the mine were doing.

He found that a horse could tug a 500-pound weight quite steadily, and came up with the idea of "horsepower": the rate at which an average horse could keep working through a long day. If he could offer a mine owner a steam engine that cost less than a horse and its feed, yet would operate at more than the rate of one horsepower, then, he realized, he had a chance of making a sale. With the spread of the metric system, his original term was replaced, most justly, by the term *watt*. Horses are strong, and the power of a single watt was defined as a small fraction—just about $1/750$th—of a single horsepower. Today's DJs don't use horses tugging on pulleys to spin their recordings or pump their speakers, but when they say they're going to power up a 750-watt sound system, they're basically saying that their equipment needs as much power as a single tugging horse could provide. A watt simply measures the power that the pushing volts and scurrying amps will provide.

In the time of Mr. Watt's descendant Robert Watson Watt, Britain's radar defense depended on being able to compute how this accurately measured power would travel through the air. The sky is loaded with electric charges, yet they're so tightly bound up with one another in air molecules that even a powerful radio wave, shooting upward, won't be able to drag them apart. Only the far stronger force fields emanating from huge storm clouds can do that, to begin the crackling of a lightning bolt. But would the weak beam of a radio broadcast be enough to tug loose at least some of the electrons in a soaring metal airplane?

That's what Arnold Wilkins had to compute in Slough in 1935, and he used Mr. Watt's ideas—and the clear definition of watts that the original definition of horsepower provided—to work out that there would, in fact, be just enough power transmitted outward to make those electrons transmit a detectable answering signal back toward the ground. Power measured in watts would fly up, a field with a pushing force measured in volts would accordingly appear around the plane, and a flow of electrons measured in amps would start up inside the airplane's metal wing—strongly enough, Wilkins computed, to create the answering signal on which Britain's radar defense could depend.

NOTES

5   *conveniently vain . . . Alessandro Volta*

Volta's discovery didn't come out of the blue, for he'd been working on devices to transfer or measure static electricity for over a quarter-century. But when Luigi Galvani—a mere anatomist—found what seemed to be a source of moving electricity, the semi-aristocratic Volta was appalled, and his research went into overdrive.

Volta is surprisingly restrained in the key paper where he announced his battery to the Royal Society, but the true stories of berating competitors, and especially the unfortunate Galvani (who got the explanation of the bimetal effect almost entirely wrong) is well described in Marcello Pera, *The Ambiguous Frog: The Galvani-Volta Controversy on Animal Electricity* (Princeton: 1992). The Fara and Heilbron texts in the reading guide for this chapter give wider background.

5   *tingling sensation . . . across his tongue*

The atoms of the two metals had different configurations of electric force on their surface, but that alone would not propel a current across Volta's tongue. Saliva, however, is largely salt water, which made it reactive enough to combine with the zinc, so that microscopic fragments from the zinc floated away into Volta's mouth. This meant that instead of balanced amounts of electric charge remaining on each disc, a buildup of raw electrons began to appear.

With those extra electrons in position on the zinc, the potential "push" between the two different metals could get to work. Sulfuric acid is even better at reacting with metal to feed electrons into one of the waiting terminals, hence its use in later batteries.

Note the crucial role of the liquid. The pressure or "oomph" between the two metals—the voltage—depends on the nature of the metals involved. That's why batteries have standard voltages such as 1.5 volts, and why Volta was safe from electrocution: his coin-shaped discs couldn't generate much voltage. But the *energy* to tear apart the zinc atoms, and supply the electrons to keep on taking that ride from one metal to the other? That comes from the sizzling liquid, lapping away around the metals—and that's why Volta would continue to feel the tingling sensation, so long as he kept the coins in his mouth.

Curiously, our civilization isn't the first to use batteries. In the 1930s, a strange, rusted, urn-like device was found in Iraq that was almost certainly a battery. It dated from the third century A.D. and had an iron rod at its center, neatly separated from a copper sheath. When twentieth-century researchers rebuilt it and put vinegar in to dissolve extra charges on the copper, they found it produced a steady one-volt output. But whether it was used simply to electroplate jewelry, or in terrifying, spark-producing priestly rituals, archaeologists cannot tell.

5   *the world's first steadily operating "battery"*

The name had first been used for groups of metal-coated glass containers, called Leyden jars, which could store accumulations of static electric charges. (A repeated grouping of identical objects is often labeled a "battery," as with a battery of artillery.) But Leyden jars could release only a single sudden shock. Volta's construction was superior because it produced a steady current.

A single pair of metals provides only a small amount of the pushing force that Volta noticed, but connecting another pair of metals to the first doubles the effect; connecting a third pair triples it. The diagram Volta attached to his Royal Society paper showed dozens of metal pairs connected by wires; a modern auto battery has similar layers. Giuliano Pancaldi's *Volta: Science and Culture in the Age of Enlightenment* (Princeton, NJ: Princeton University Press, 2003), 246–48, shows the gradual domination of the word *battery* over competing phrases such as *trough* or *cell* (the latter of which lingers on in our term *fuel cell*).

7   *a primitive mobile phone . . . in 1879*

The creator was David Hughes, an American engineer living in London, and he used a trolley to carry the "mobile" phone, which created audible clicks when it picked up sparks that were generated up to five hundred yards away. There's a copy of the original at London's Science Museum.

13   *that filled empty space*

You can watch the Big Bang on television now. Only a very small portion of the radiation seething through space in the universe's earliest moments was used up in creating the charged particles from which we're built. Most of it kept flying loose, and since it's simply electromagnetic radiation, similar to that of television broadcasts, we pick up some of it whenever a TV set is tuned between stations—creating perhaps 1 to 5 percent of the snow-like static that fills the screen.

23   *judicious financial involvement . . . with Congress*
Morse secretly hired the chairman of the House Commerce Committee, Francis O. J. Smith of Maine, promising him substantial profits if the committee were given the needed report. Shortly after the committee gave its favorable report, Smith retired and took up a position as Morse's partner. He got rich from Morse's patents, and the government funds he'd just voted for, then left Morse's employ and tried to get richer by blackmailing Morse.

34   *In 1875, his . . . invention came together*
Many other inventors were active in the field, not least the unfortunate Elisha Gray, who filed for a patent just hours after Bell. Gray wasn't too perturbed, though, for he suspected that the speaking machine was just a distraction from his main research on telegraphs. Even Bell was confused, at least for a while, and explained to investors that what he'd created was merely a telegraphic machine that didn't need experts to translate its signals.

36   *but that's basically how a telephone works*
The ringing we hear when we phone someone doesn't come from the telephone of the person we're calling. Rather, it's a signal being sent to us from a central switching station to give us the *impression*

that we're hearing that phone. It's a trick that dates from the early days of central exchanges. (If the two signals fall out of sync, the illusion vanishes; this is when the person at the other end of the phone hears a ring and answers before we've heard a sound.)

41 *Edison thought about it and saw there*

Edison was an unscrupulous man but a deft engineer. The way he modified the battery's current was by putting an obstacle in its way: a tiny box, filled with powdery granules of carbon.

When someone speaks loudly into such a telephone, those carbon granules are blown tight against one another. But when carbon is jammed together like this, electric currents travel across it more easily—think of how much easier it is to advance across steppingstones that are suddenly brought closer together. Only as the human voice goes back to being softer do the carbon granules stop being squeezed so tight. The carbon in the small box is powdery and loose again, and so less of the battery's current makes it across. His invention continued to be used for almost a century.

53 *"I had no time to do anything more about it"*

And if he hadn't been so busy, he quite possibly would have invented the television. The black dot that Edison saw came from streams of electrons boiling out from his lightbulb's filament. He'd already found he could partly control the beam by putting a bit of tinfoil on the outside of the glass, and if he'd gone ahead and put magnets beside the bulb, he would have found the electrons being tugged from side to side by the magnet's field; this was the essence of J. J. Thomson's subsequent experiment. Even with his limited measuring techniques, Edison would have noticed the black dot appearing in different places; had he tried different coatings on the glass, he would have seen different colors appear and glow.

That's almost exactly how traditional cathode-ray television sets or computer monitors work. A beam of electrons is shot from the back of a tube that's basically a big lightbulb; when it hits sensitive molecules coated to the inside of the glass front of the tube, the energy the electrons carry makes those molecules glow. To steer the electrons and create a moving picture rather than just a single heap of dots, magnets on either side of the tube pull the electron beam from side to side, neatly matching signals that the broadcaster is sending out.

55   *the sparks caused by static electricity*
Static electricity powered the atomic bombs over Japan.

In most atoms, the powerful electric charges inside a nucleus can't push away from one another. That's because what's termed the strong nuclear force acts like a powerful glue, holding them together. But the nuclear force is only about one hundred times as strong as the force of electricity. As an atom gets close to having one hundred protons in its nucleus, that glue is easily dissolved. Uranium and plutonium atoms are about that size, and so relatively easy to split apart.

In the bomb exploded over Hiroshima in 1945, charges in the uranium nuclei that had been held together for billions of years were abruptly released. For a very brief moment, there was nothing to hold back the electrostatic repulsion between the protons, and an entire city was flattened as they flew apart.

66   *an invisible force that spread from a moving magnet*

The final details are modern, but the vision of invisible hovering forces is old. William Gilbert, the court physician to Queen Elizabeth I, wrote that the reason rubbed amber could attract feathers was that a "humour" was being removed from the amber, which led to a powerful "effluvium" floating around it. The words seem odd, but substitute the terms *charge* and *field*, and he would then be saying that charge is removed from amber when it's rubbed, and the result is a changed field around it—which we would take as a rather perspicacious suggestion. For Faraday's own changing views of what constituted a field, see the guide to further reading for chapter 4.

68   *his excited theories . . . were politely set aside*

Or, sometimes, less than politely. The *Athenaeum* magazine wrote that Faraday should go back and study his school mathematics before venturing into the deep seas of modern physics; the immensely supercilious Astronomer Royal, Sir George Biddell Airy, remarked that "I can hardly imagine anyone who practically and numerically knows [modern electrical theory to accept] anything so vague and varying as lines of force."

Airy made a habit of putting down his social inferiors. Aside from being a footnote in the history books for missing Faraday's work, he's also a footnote for missing the discovery of Neptune—once again, a scientist from the wrong background presented him with evidence of its predicted orbit, which he refused to examine properly.

76   *it would also writhe sideways*

In the language of electrical engineers, the different layers of the cable and the saltwater built up a series of "capacitors"—that is, there were two separated conducting surfaces, which both tried to pull some of the current and charge. That was what produced the distortion of the signal, for when so much of the loaded current was diverted into the cable's iron skin, only a weak and spread-out signal could get through the main copper core.

85   *only very* gentle *battery pressure could be applied*

By the end of the project, Thomson was powering telegraph messages across the Atlantic using only a thimble and a single drop of sulfuric acid. He would lower two different slivers of metal into the thimble, let the acid react with one of them to create a buildup of electrons (just as Volta's saliva had reacted with the coins on his tongue), and then connect that to the giant undersea cable. The sliver of metal was tiny, but electrons are tinier still; many billions of extra electrons quickly accumulated on the sizzling metal. From that skimpy "pile" of charged electrons, a force field powerful enough to cross the ocean streaked forward, jostling waiting electrons in a telegraph receiver two thousand miles away.

87   *just a force field waiting at the sockets*

The idea of fields is fundamental, but the terminology of "voltage" —to describe the force or "oomph" with which the electrons in a current are pushed—is immensely convenient. The appendix on Mr. Volt touches on this, but to understand the link in some more detail, consider again what happens when you stretch your fingertip close to a metal railing on a dry day after you've been unwisely scuffing your feet. The air between your finger and the metal is ordinarily an insulator, for there are hardly any free electrons available to carry a current in ordinary air. If you keep your finger several inches away from the metal you're not going to get a shock, but if you bring your finger closer to the metal, you're doing work, from all that effort you've spent pushing the charges on your finger closer to the gleaming metal danger. The field is more intense around your finger now, for it's concentrated in the narrower gap between you and the close metal.

Suppose you snipped off the extra charge from your finger at that point, and left it hovering in position, that fraction of an inch from the metal railing. Since the field is strong here, it would send the charge shooting back to where you'd been. A "voltage" measures that pushing force, the difference in what potentially might happen.

If the wire is in an area where the field's not changing, there's a curious safety. That is why a bird can safely sit on a high-voltage power line. When both its feet are on wires of the same voltage, there's no difference between the pull that the associated force fields can give to the electrons in its body. Let the bird extend one foot to an aluminum ladder that's touching the ground, however, and a sudden explosion of coq flambé will result.

87   *they drift along so slowly, barely at walking speed*
Individual electrons travel fast, but since they also bounce in all directions they make hardly any net progress forward. Only if a wire could be cut open and the electrons somehow released

would we see their intrinsic speed, for many would be going fast enough to shoot straight out of the atmosphere and into outer space. What the force field from your wall socket does inside a wire is make sure that the majority of the high–speed crashes jostle forward in the same direction.

90   *Whenever you shake that charged-up finger*
We create and launch electromagnetic waves all the time. Comb your hair on a dry day and you tear a small number of electrons from your hair. They accumulate on the comb, and if you then wave the comb in a graceful andante, taking five seconds to swing the charged comb to the right, and another five seconds to swing it back to the start, you will have pumped out a wave whose front has a head start of ten seconds on its back. The wave travels at 186,000 miles each second, so in the ten seconds of this graceful comb–conducting you've created a soaring wave that stretches upward into outer space and is 1,860,000 miles long.

Other launch devices are also to be found around the home. If you wash your dog on a dry day, and towel it vigorously, there's likely to be a slight buildup of static charge on its tail. As the dog happily wags its tail, creating a complete back–and–forth cycle in just half a second, then it's creating an electromagnetic wave that stretches 93,000 miles (i.e., half of 186,000) from head to end. This invisible hurtling wave will reach the moon in less than two seconds; it will reach the orbit of Saturn in a bit under an hour, and the canine–launched beacon will exit the solar system a few hours later, and keep on going.

90   just by jostling an electric charge

If shaking any charged particle can send out these waves, why doesn't the movement of electrons inside a computer send out similar waves? The answer is that it does: not very powerfully, and at highly variable frequencies, but it's one of the reasons that airlines have stewardesses check that no one's using a laptop at the delicate moments of takeoff and landing. The undulating fields that shoot out from laptops or other equipment this way—and especially the undulating fields from the active processors inside such devices—could fly the length of the plane and easily bounce off the wing, affecting the pilot's controls.

114   His name was Robert Watson Watt

But not for long. To the distress of archivists everywhere, after Watt was knighted in 1942, he changed his last name, and became the double-barreled Watson–Watt. Frederic Lindemann engaged in yet another of these peculiarly British morphs; in his case, through the help of Winston Churchill, he became the impressively named Lord Cherwell during the war. For ease of reference, I've stuck to the earlier names throughout.

118   Plutos were floating loose

The electron gas model of metals stemmed from the work of Paul Drude, working at Leipzig in 1900. It was long out of date by the time Wilkins and Watson Watt used it, and they knew it was "wrong"—atoms aren't little solar systems, and electrons aren't little planets.

But they could still use it, for Drude's theory was not so much false as just incomplete. This happens a lot in physics. A larger domain is seen, in which the previous theory is realized to be a special case. But so long as you stay within the limits of the smaller domain, the archaic equations and vision that came from that domain will still accurately apply. Newton's laws of motions, for example, are still good enough for our ordinary environment, even though they're actually just a special case of Einstein's far richer special theory of relativity. They give valid results, so long as one stays away from the realm of speeds approaching that of light. In a similar way, Mozart's simple triads can be used by modern composers, even though Debussy and then, even more, the American jazz greats opened up a range of possible chords beyond what Mozart ever imagined.

For the origins of Drude's model in the kinetic theory of gases, see Walter Kaiser, "Electron Gas Theory of Metals: Free Electrons in Bulk Matter," in *Histories of the Electron: The Birth of Microphysics*, edited by Jed Z. Buchwald and Andrew Warwick (Cambridge, MA: MIT, 2001), 255–303.

119   *by pumping up ... radio waves ... the enemy plane [could] become a flying transmission station*
Which is why mirrors work. Ordinary light is a wave much like those transmitted from Britain Chain Home radar towers, only with a shorter and more energetic wavelength. When a light beam made of these waves hits the metal coating at the back of a piece of glass, loose electrons there begin to shake.

Like all jostled electric charges, those electrons start broadcasting out their own Faraday–style waves. If the metal backing of the glass is jagged and rough, the broadcasts fly out in all directions and we just see a dark blur. But if the back of the mirror is very smooth, the broadcasts lift off in stately flight patterns beside each other, carrying a close duplicate of the original image that flew in. Smile in a mirror, and multitudes of mini-radar broadcasts, lifting out from the ancient metal atoms back there, are what you see smiling back.

123 *the less revealing label "Radio Direction Finding"*

"When Rowe and I sat down to devise a label for the system . . . we said to ourselves "Let's think up something that doesn't merely conceal the truth but positively suggests the false. . . . We agreed on the initials "R.D.F." [for radio direction finding]. . . ." Watson Watt, *The Pulse of Radar*, 123.

It's possible he and Rowe could have chosen a worse code name, but it would have taken some doing. When they came up with the cover name, they were pretty sure their oscilloscopes would never be good enough to get a clear bearing on an airplane's direction. Soon after they selected it, though, improved technology meant that British stations did begin to achieve accurate direction findings. Their secret label had became a blaringly informative clue.

138 *The Würzburg radar . . . poured out waves of a bare ten inches*

A close descendant of the Würzburgs sits in our kitchens today, for when the wave generated by a radar transmitter is just a little bit shorter—two or three inches is ideal—that radar beam will cause any water molecules it hits to vibrate. A few postwar researchers thought this process might be useful for curing tire rubber, but other markets beckoned. Waves of three inches in length are known as microwaves—and so the microwave oven was born.

Since the oven is still basically a radar transmitter, any loose electrons it hits get tugged back and forth with such power that they can fly loose and even start sparking. This is why putting metals—which are laden with extra electrons—inside the machines is not recommended.

144   *no one else there dared to speak in his support*

Why did so many officers go along with Harris? Partly it was because precision bombing hadn't worked; in late 1941, for example, only one fifth of all sorties managed to place bombs within seventy-five square miles of the target. Partly it was the great momentum of having created all those bomber squadrons, and airplane factories, and trained aircrew—for who could justify not trying to use them just a little bit more? Left unspoken were the memories of the World War I trenches, which Harris had flown over, and the strong feeling that any aerial attack was better than sending British troops into land battle on the continent of Europe again.

When those arguments missed—and what exasperated the Royal Navy and other branches of the British armed services so much—is that resources that went into bombing could not be used for anything else. A vast amount of Britain's wartime GNP went into Bomber Command, and so was not available for extra destroyers, artillery, transport aircraft, and everything else.

146   *Some of the ground controllers in such circumstances yelled . . .*
"Break off . . ."
The recording is from attacks on Essen the following night. No
transcripts or recordings of ground radar staff survive from the
Hamburg night. See David Pritchard, *The Radar War* (Wellingbor-
ough, Northamptonshire, England: P. Stephens, 1989), 213.

148   *"They were on their hands and knees, screaming"*
German civilians weren't done with terror once the bombers left.
After a similar giant raid on the city of Cologne, also led by Harris,
the Nazi government made all surviviors sign the following
pledge: "I am aware that one individual alone can form no com-
prehensive idea of the events in Cologne. One usually exaggerates
one's own experiences and the judgment of those who have been
bombed is impaired. I am therefore aware that reports of individual suffering can only do harm, and I will keep silence. I know
what the consequences of breaking this undertaking will be."
From *Dresden: Tuesday 13 February 1945*, by Frederick Taylor (London:
Bloomsbury, 2004), p. 128.

149   *the world . . . with teleporting jumps*
The term "teleporting" is frequently used in describing the quantum realm. But it does suggest that the objects so described have a
continuous identity before and after their "leaps"—and a central
tenet of quantum mechanics is that such identifications are inher-
ently impossible at the level we're discussing.
    The concepts can be confusing even to specialists, and the
reading guide keyed to this section might give some solace to the
perplexed.

153   an exclusion zone seemed to operate

This is Pauli's Exclusion Principle, and the effect is not so much on linear speed as on the overall energy that the electrons possess. For the strange thing about electrons is that two electrons can't have the same energy, just as two people can't occupy the same point in space. If one electron does already occupy a particular energy state, it can actually stop another electron from moving up into it—much like blocking a particular rung on a ladder.

Pauli's principle is immensely powerful, for atoms are almost entirely hollow, and without this restriction on the way electrons can overlap, we'd be in trouble. Even when you just hit your fingertips to the table in a vainly time-wasting drum roll, the great empty spaces in the atoms of your finger would travel right through the great empty space in the atoms of the table if it weren't for the Pauli principle.

Your feet would start sinking through the floor, just as your posterior would tumble through the chair. There might be a brief moment of reappearance—as your body fell into the airspace of any room below—but when you reached the floor of that room, you'd simply sift through again, a process that would be continued to distressingly great depths in the planet below.

That's why our whole life is, safely, spent hovering. We hover over the floor when we walk, and we hover over a chair when we sit. Even a dedicated couch potato, slumped as slothfully as possible on the sofa before a television, is also being propped up in the air by the wonders of quantum mechanics: his body kept floating by the resistance of his electrons to sharing too many energy states with those of the couch.

Charles P. Enz's *No Time to be Brief: A Scientific Biography of*

Wolfgang Pauli (Oxford, England: Oxford University Press, 2002) is excellent for background on Pauli. For more on the associated topic of the virtual electrons that "shoot up" from electrons and help preserve our separations, see Richard Feynman, *QED: The Strange Theory of Light and Matter* (Princeton, NJ: Princeton University Press, 1985); for Pauli's at times maddening role in the formative years of QED, along with Enz, see Silvan Schweber, *QED and the Men Who Made It* (Princeton, NJ: Princeton University Press, 1994).

156   *and in 1920s England . . . at its private schools*
"The great thing about a [private] school education," Turing said, "is that afterwards, however miserable you are, you know it can never be quite so bad again." (What Americans call private school, the British call public school.) The quote is in Andrew Hodges, *Alan Turing: The Enigma* (London: Vintage, 1993).

161   *What he needed . . . were the new insights in physics*
Until then, Turing and all other researchers had to make do with the intermediary technology of vacuum tubes (the "valves" of British terminology). These were basically miniaturized lightbulbs, with added wires or metallic meshes inside that attracted the electrons flying up from a heated filament and accelerated them upward. This was good for making weak signals stronger, but vacuum tubes were miserable devices to work with.

Cold filaments don't boil out many electrons, so there was a long wait while the tubes heated up (which is why old electric devices needed "warming up"). The glass around the hot filaments had to be airtight, which meant the tubes regularly overheated;

the tiny filaments also cracked. Anyone using large numbers of vacuum tubes knew to keep equally large boxes of spares handy. As John Pierce—who later coined the label *transistor*—put it, "Nature abhors a vacuum tube."

162   the British government's codebreaking group

Actually a "cipher-breaking" group. A code is a direct-substitution system, where one word is used to stand for another, as with a child's code in which, for example, the words *national embarrassment* are substituted wherever the word *shrub* has been written. A cipher involves more-complex substitutions of the components, ranging from Julius Caesar's early efforts, where each letter of the Latin alphabet was replaced by a letter three places later in the alphabet, to far more intricate ones, where complex intermeshing gears produce the substitutions. For simplicity I follow the informal usage, where the term *code* is used to cover both types.

There's a nice twist here. Without radio, there wouldn't have been Bletchley Park, for before radio, military messages were sent directly to their intended targets, and outsiders couldn't listen in simply by setting up a big antenna. But without Bletchley Park, when would there have been computers? Such are the unexpected ways in which electricity fed its own further development—radio begetting codebreaking centers, and those centers helping beget computers.

166   The machine was more than a calculator

The first Colossus in 1943 could quickly test a possible solution against an encoded text, but then had to be reset by hand. An improved version was soon built at Bletchley—wartime pressures concentrated the minds of procurement directors wonderfully—and this one could change the target decodings it was considering without being reset from outside. But although this meant there now existed machines that could automatically make choices, it still wasn't fully programmable, nor could it store programs.

Turing was only marginally involved with its construction, but Colossus was initiated in great part by Max Newman, who'd been one of his lecturers at Cambridge, and Turing was kept informed of its workings.

169   a new technology being perfected in America
Although the transistor technology came from America, the logical switching it would follow built on work of the mid–nineteenth-century English mathematician George Boole, who took it upon himself to codify every logical thought that could possibly exist. Such an obsession would not be considered too odd in certain polite English circles even today, but what's curious is that he succeeded.

Boole wrote down his results as simple equations, which is where his connection with computers comes in. For if any two true statements are combined, then the result is also true, and he'd write that as $T+T=T$. If a true and a false statement are combined, then the result is false—he'd write that as $T+F=F$.

It seems a bizarre, elaborately obvious arithmetic, of the sort only Lewis Carroll would appreciate, and it became even more bizarre when Boole wrote "1" for "True" and "0" for "False." In that system, the previous equations become $1+1=1$ and $1+0=0$. This is binary code.

Logicians liked this, the ordinary world ignored it, but by 1937

Claude Shannon had realized that relay switches controlled by electromagnets could actually carry out these equations Boole had proposed so long before. The transistors created a decade later could do it even better, for a solitary signal coming into a switched-off transistor won't be enough to get a current moving; at most that signal will jolt the silicon into its active, current-carrying state so that a *second* signal can now pass through without hindrance. In other words, it takes two signals coming into such a transistor to get one signal going out. But if the number "1" is taken to represent a signal, this is just saying that inside a transistor, $1+1=1$. If the absence of a signal is written as "0," then a transistor will also operate to ensure that $1+0=0$.

What had happened was wonderful. Boole had looked inside the human mind and pulled out his odd-looking equations about how truths and falsehoods operate. Transistors can imitate those equations, but this time inside solid rock. Thus did one obscure nineteenth-century mathematician pave the way for our inner thoughts to be accurately copied inside silicon grains. For Boole's original work, seé George Boole, *An Investigation of the Laws of Thought* (London: Dover Publications, 1995). For more on Shannon, see the chapter "Understanding Information, Bit by Bit," in *It Must Be Beautiful: Great Equations of Modern Science*, edited by Graham Farmelo (London and New York: Granta Books, 2002).

177   *intruder atoms such as phosphorus*
What Bardeen and Brattain actually found when they inserted the phosphorus into their semiconductors in late 1947 was that instead of making negative charges race forward, they apparently were making positive charges race backward.

Bardeen was puzzled—"this is the opposite of what one might expect," he wrote in his lab notebook—but engineers are good at following up what works. He and Brattain soon realized that instead of inserting an atom with extra electrons to carry a current, they had inserted atoms that had *fewer* electrons than the others around them. Their perfect lattice now had holes in it: gaps where those new atoms had an unfilled outer shell.

Electrons from the other atoms started falling into those holes, but each time an electron fell forward, it left a fresh gap where *it* had been. Other electrons rushed forward to fill those gaps, which meant there were now free spaces even farther back in the rocky lattice. A bizarre empty hole was speeding backward, inside a solid crystal. It was a roundabout approach, but they had found a way to fling holes from one point in a solid material all the way through it to the other side. It wasn't the controlled transfer of negative electrons they'd first expected, but it worked. (The actual experiments were with germanium, not silicon: both elements have four "gaps" in their outermost shell, but germanium is slightly easier to work with.)

177   *But Ohl . . . knew enough of quantum mechanics*
The extra insight that the quantum engineers used was that electrons were as much like waves as like particles. This means that the silicon or germanium crystals are suffused with electron waves, which can interfere with one another across the whole volume of the crystal. The insights are summarized in the band theory of solids. Rather than looking at the individual electrons around one atom, and asking whether they are tightly attached to the atom or available to be pulled elsewhere in the solid, one

instead looks at the assemblage of all electrons in the solid and measures analogous properties for them as a whole.

Watson Watt's colleague Arnold Wilkins knew that the summated activities of great numbers of electrons would approximate to distinct conduction bands, and this is why Drude's pre-quantum image of individual electrons being those conductors was valid enough for his calculations. It's also why Bardeen and Brattain could talk about conducting holes, without however having to imagine that the hole is literally being passed from one atom to the next. The Al-Khalili or Polkinghorne texts in the reading guide for this chapter give good background.

178   *new chemical fabrication techniques*

Generations of engineering and computer professors have sighed over the way their students place a flowering "geranium" at the heart of transistor technology. To keep the spelling straight, it helps to remember the context in which the element was discovered. There was bitter hatred between France and Germany after the Franco-Prussian War of 1870–71, so when the French chemist de Boisbaudran discovered in 1875 a new element that Mendeleyeev had predicted, he named it *gallium*, after the Latin name for France. When the German researcher Clemens Alexander Winkler, a decade later, discovered yet another element Mendeleyeev had predicted—one that occupied a gap directly beneath silicon in Mendelyeev's table, and so had many properties in common with silicon—there was no question but that he would name it *germanium*, after his newly triumphant country.

179   the rock itself doesn't have to move, swiveling . . . like a big metal switch

What's being described is simply the idea behind a standard three-layer transistor. The outer layers are configured (though having an excess of electrons) to easily carry negative charge. The middle layer blocks them. But since that middle layer is a *semi-conductor*, its nature easily changes. Even a very slight increase in a current that's fed in will make it transform, so that the charges from the outer layers can get across.

Note the similarity to Edison's carbon-button microphone. In a modern hearing aid a battery is constantly attempting to push a steady current across the middle layer of a transistor, but only the weak voice signal fed in from outside will make that middle layer transform to let the current across. The stronger the voice, the stronger the current from the battery that gets through.

If the effect was exactly one for one, the transistor wouldn't be much help, but the great power of transistors comes from the fact that the middle layer is so precarious that even a slight change in the input—the slightest shift in the voice to be heard—can make it transform considerably. There will be an enormous "power gain," and it was measuring a similar power gain in November 1947 that made Brattain and Bardeen realize they were on the right track.

181   Hopper liked explaining in later years

She also liked describing the day in 1947 when she found a moth shorting the circuitry in the computer she was working on at Harvard. "First actual case of a bug being found," she wrote next to its carefully preserved carcass, which she taped in her log-book. The term seems to have been used as a generic explanation for mysterious faults in electric circuits from time to time as far back as Edison, but with Hopper's central position at Harvard, plus the taped-up evidence, the label "bug" for a computer fault now took off.

182   an electronics magazine came to take a photo
It's the photo used in innumerable textbooks and histories, taken for the September 1948 cover of *Electronics*, but it has as much accuracy as the ones emanating from the Kremlin in its Politburo days. Walter Brattain can be seen directly behind Shockley, who's peering aimlessly through Brattain's own microscope. When Brattain was young, he'd once spent almost a full year in the mountains, largely on horseback, guarding a cattle herd with a rifle on his lap. In the photo one can see his tensed hands, tilted slightly forward, in a path that, if continued, would let him wring Shockley's neck. Forty-five years later Bardeen was interviewed about that day. "Boy, Walter sure hates this picture," the mild Bardeen explained. See Michael Riordan's and Lillian Moddeson's *Crystal Fire: The Invention of the Transistor and the Birth of the Information Age* (New York: W. W. Norton, 1997), p. 167.

182   Bardeen left . . . no transistors were mass-produced in time
Bardeen and Brattain were surprisingly close to making the great breakthrough to the MOS (metal-oxide-silicon) transistor, which became the heart of Intel's chips in later decades. Although their final 1947 experiments had used germanium, from which the key oxide layer needed for MOS transistors washed off, if Shockley had let them continue, they would probably soon have turned

back to silicon—which they'd been following in great detail—and on silicon the crucial oxide layer would have remained.

188   The incoming wave, feeble as it is
Battery power is at a premium aboard satellites, and the antenna on a GPS satellite glows with the energy of only five small light-bulbs. By the time the signal reaches Earth's surface, having had to push through the thick atmosphere and having dispersed over thousands of miles, our receivers have less than one-billionth of a watt of power to work with. The transistors in our GPS receivers are sensitive indeed: an ordinary toaster needs about one trillion times more power than that to warm a single piece of bread.

200   many people had seen demonstrations . . . The twenty-year-old Mary Shelley
She hadn't seen the most famous demonstration herself, for she'd only been five years old in 1802, when Galvani's nephew, Giovanni Aldini, had come to London and obtained permission to set up his equipment around an understandably worried young convict, Thomas Forster, who was very soon to become a nice fresh corpse.

Forster was lifted up, hanged, then his body brought down; Aldini poured paste into Forster's nostrils and mouth, and then connected a battery. For a horrible long moment Forster's body seemed to live again, and "convulsions appeared to be much in-creased," appearing in his "head, face, and neck, as far as the deltoid."

The gruesome effect was much discussed, not least by Percy

Shelley—Mary's future husband—who as a schoolboy often used voltaic piles to charge up his body till his hair stood on end. See Esther Schor, ed., *The Cambridge Companion to Mary Shelley* (Cambridge, England: Cambridge University Press, 2003).

202 ¸ *Any atom that has a different number of electrons . . . is called an* ion

When we talk about the "pH" of a solution, that's simply shorthand for counting the electrically powerful ions inside it. In a glass of water there's one bare hydrogen ion on the loose for about every 10,000,000 ordinary molecules, which is why water's pH is said to be seven—the number of zeros in 10,000,000. The hydrochloric acid in our stomach has one hydrogen ion for about every hundred water molecules, and so had a pH of about two. That greater density of electrically charged ions attacks the bacteria on the food we swallow, and it sizzles into the cell membranes of the food itself.

Cosmetics companies often advertise their products as being pH balanced, which means they have a pH of seven. The labels are less forthcoming in explaining how this expensive feat is achieved. Quite often, the customer is simply buying a product mixed with a great deal of liquid that has those hydrogen ions diluted to one in 10,000,000. In other words, the customer is spending most of his money on water.

206 *When we have a thought, and a nerve cell in our brain begins to fire*

Since the simple movement of electrons inside a wire sends out radio waves, might not the natural shaking of electrons inside the human brain send out similar invisible waves? Almost as soon as Hertz's experiments became known, a number of researchers, led by the radio pioneer Oliver Lodge, thought that this could be the scientific basis for extrasensory perception (ESP). In the 1920s, the spread of ordinary radios—little boxes that really could detect invisible messages from afar—bolstered the general public's belief in phenomena like ESP.

It turns out that although the brain does in fact generate low-power radio waves, they're too weak to be clearly detected at any significant distance. The explanation has to do with the trade-off between the length of a wave and the number of those waves it takes to cross a given distance. Hodgkin and Huxley confirmed that the nerve cells in our brain fire at the fairly slow rate of a few thousand times a second, so as with the comb calculation on page 246, this means the front of each wave has a head start of about one-thousandth of a second on the back of the same wave.

A bullet can't go far in one-thousandth of a second, but it takes only one full second for electromagnetic waves to travel 186,000 miles, which means they'll travel 186 miles in one-thousandth of a second. That's what emerges from our brains, and at first this fact—that invisible waves with a distance from peak to peak of 186 miles are constantly pouring from our heads—seems to be solid evidence for extrasensory phenomena. (The waves can be a little longer or a little shorter, depending on the actual rate of neuron firing.)

But it's not quite as good as it sounds, for 186 miles is a huge wave. Cell phones produce electromagnetic waves with a length of only a few inches; even AM radio produces waves with a length of only a few hundred yards. For a wave to be easily detected, it needs to be produced from a source that's relatively large compared to the wave. But the human head is tiny compared to 186 miles. That means the wave is generated very inefficiently, and so the field it produces is far too weak for humans to detect unaided; the fact that the signals interfere with one another makes it even feebler.

210   *Sigmund Freud . . . plant extract called cocaine*
Freud almost got credit for discovering cocaine's anesthetic properties, but he got distracted by other work and didn't take the time to elaborate his findings. Instead of blaming himself for this professional failure, however, he wrote: "looking back . . . it was my fiancée's fault if I did not become famous in those early years." But he didn't hold a grudge, and more than once sent her small vials of cocaine to sample; he even began taking it regularly, on and off for almost a decade: it helped him relax, and—as he promised her before another visit—it also turned him into "a big wild man with cocaine in his body." See Peter Gay, *Freud: A Life for Our Time* (London: J. J. Dent & Sons, 1988), 42–45.

211   *The mechanism was exactly the same as in humans*
Although humans need sensitive lab equipment to detect the electric currents in live nerves, many animals don't. The seemingly cuddly duck-billed platypus, for example, hunts in muddy river bottoms at night. The crayfish or shrimp that are its prey try to hide in the mud, but their nerves—just like ours—are constantly pouring charged sodium ions back and forth. Those moving charges send electromagnetic fields undulating outward. The platypus's beak is loaded with cells that can detect those fields;

with a quick bite, or an accurate slash from its venom-filled ankle spurs, the crayfish is gone.

Hammerhead sharks are even better at this, for the vast space in the "hammer" swelling on their heads means they carry even more cells that can detect electric fields. The shark's prey might be cowering unseen around a turn, or burrowed deep into layers of sand. But that prey's heart is going to be beating, and the quivering pulses of its heart muscle are controlled—again, like ours—by microscopic pumps in the walls of its nerve fibers, pouring charged sodium and other ions back and forth. The invisible electromagnetic fields stretching outward begins to pulse. The hammerhead detects that, even in total darkness, and turns closer, jaws widening, to . . . investigate.

218   the future that insights into neurotransmitters will create still remains to be seen

What happens to personal responsibility when we've found a biological source for all our actions? "If we can find explanations for the evil people do, then are we not replacing moral evil, which is freely undertaken, with natural evil, which is beyond our control? To the volcano and the virus might we now add the dysfunctional amygdala and the abnormal orbitofrontal cortex?" (Sean Spence, University of Sheffield, in *New Scientist*, March 20, 2004.)

220   as many . . . as there are stars in the Milky Way galaxy

Kant wrote that "Two things fill the mind with ever new and increasing wonder and awe—the starry heavens above me and the moral law within me."

He was more right than he could have known. The numerical

similarity is just chance, but quantum fluctuations in the very early universe seem to have been responsible for the distribution of galactic clusters that we find in the starry heavens above; those same quantum fluctuations control neuronal processing of each human brain contemplating the moral law within.

For a one-sided yet entertaining discussion of free will and quantum mechanics, see Roger Penrose, *The Emperor's New Clothes: Concerning Computers, Minds, and the Laws of Physics* (Oxford: Oxford University Press, 1989).

GUIDE TO FURTHER READING

EARLY ELECTRICITY

The classic account of electricity's early years is John Heilbron's *Electricity in the Seventeenth and Eighteenth Centuries: A Study of Early Modern Physics* (Berkeley: University of California Press, 1979). This has such vignettes as the Englishman Robbert Symmer's baffled report, in November 1758, that "I had for some time observed, that upon pulling off my stockings in an evening they frequently made a crackling or snapping noise; and in the dark I could perceive them to emit sparks of fire"; Thomas Hankins's *Science and the Enlightenment* (Cambridge: Cambridge University Press, 1985) homes in on the way eighteenth-century investigators tried exploring the observations of Symmer and others, through such expedients as leading the static crackling along long cotton strings, creating something like simple telegraphs; he leads up to the seminal experiments of Galvani and Volta, a bickering, sulk-ridden relationship described with much delight in *The Ambiguous Frog: The Galvani-Volta Controversy on Animal Electricity*, by Marcello Pera (Princeton, NJ: Princeton University Press, 1992).

Benjamin Franklin's contributions were slighted in this book, but Walter Isaacson's *Benjamin Franklin: An American Life* (New York: Simon & Schuster, 2003) provides an eloquent introduction, as well as evaluating whether the famous kite experiment ever happened (he votes yes). I. Bernard Cohen's writings on Franklin began with *Franklin and Newton* (Philadelphia: American Philosophical Society, 1956) and were still going strong forty years later with his *Science and the Founding Fathers: Science in the Political Thought of Jefferson, Franklin, Adams and Madison* (New York: W. W. Norton, 1995), which shows a profundity of analytic intelligence among America's political leaders not quite matched by more recent occupants of Thomas Jefferson's office. Patricia Fara's long series of writings on early magnetism are well represented in "'A treasure of hidden vertues': the attraction of magnetic marketing," in *The British Journal for the History of Science* 28 (1995): 5–35.

HENRY AND MORSE

Joseph Henry was an easygoing, matter-of-fact man, who found an easygoing, matter-of-fact biographer in Thomas Coulson. His *Joseph Henry: His Life and Work* (Princeton, NJ: Princeton University Press, 1950) captures Henry's life, down even to the way that Henry's low salary meant he had to economize on purchases of zinc—a powerful inspiration for devising electromagnets that could work with batteries powered by only the slightest slivers of that expensive metal. I'd also recommend a skim through Nathan Reingold's edited collection, *Science in Nineteenth-Century America: A Documentary History* (London: Macmillan, 1966). This has extensive extracts from Henry's letters, from the difficult days teaching ("my duties in the Academy are not well suited to my taste. I am engaged . . . in the drudgery of instructing a class of sixty boys") to his growing showmanship in using the small battery to lift a 750-pound weight.

H. J. Habakkuk's *American and British Technology in the Nineteenth Century: The Search for Labour-Saving Inventions* (Cambridge, England: Cambridge University Press, 1967) is a specialist economic history showing why labor scarcity alone does not lead to innovation. A more hands-on view comes from one young Frenchman who actually visited Albany in 1831 (when Henry was still there) and wrote up his thoughts to a somewhat wider acclaim: Alexis de Tocqueville's *Democracy in America* (many editions) is a masterpiece; see particularly part I of volume 2, on the interrelations among innovation, careers, and daily life in Henry's society.

The title of *The American Leonardo: A Life of Samuel F. B. Morse*, by Carleton Mabee (New York: Knopf, 1944) suggests it's going to be just another book of praise, but it was one of the first serious accounts to debunk Morse—his section showing Morse painfully failing to grasp the idea behind what became labeled the "Morse code" is a nice touch. Pauli Staiti's *Samuel F. B. Morse* (Cambridge and New York: Cambridge University Press, 1989) puts Morse's work in the context of his artistic efforts, showing how Morse's failure to evangelize America through his art naturally led to his attempting to evangelize through telegraphy. Kenneth Silverman's *Lightning Man: The Accursed Life of Samuel F. B. Morse* (New York: Knopf, 2003) is a good recent account; the hagiography by Morse's son (*Samuel F. B. Morse, His Letters and Journals*, edited and supplemented by Edward Lind Morse [Boston and New York: Houghton Mifflin, 1914]) shows a touching bewilderment as the son tries to explain his father's innocence over the fire that destroyed key court documents, the payoff to Congressman Smith ("[my father] was unfortunately not a keen judge of men"), and all the other misadventures in Morse's long life. Richard Hofstadter's classic *The Paranoid Style in American Politics* (London: Cape, 1966) situates Morse's mania in rich company.

On changing conceptions of time, two classics to enjoy are David Landes's *Revolution in Time: Clocks and the Making of the Modern World* (Cambridge, MA: Harvard University Press, 1983), which takes a sprightly historical approach, and Ernst Cassirer's *An Essay on Man: An Introduction to a Philosophy of Human Culture* (New Haven: Yale University Press, 1945), rooted in Kantian philosophy, which takes a more discursive line. E. P. Thompson's *The Making of the English Working Class* (London: Gollancz, 1963) excels on the religious views behind our shift to intense factory–controlled clock time; Mircea Eliade's *The Myth of the Eternal Return* (London and New York: Routledge & Kegan Paul, 1955) concentrates more on the universal principles of how societies shape time. David Paul Nickle's *Under the Wire: How the Telegraph Changed Diplomacy* (Cambridge, MA: Harvard University Press, 2003) shows, most plausibly, how the frantic haste that the telegraph encouraged led to the destabilizing diplomacy so catastrophic in the nineteenth and early twentieth centuries. Tom Standage's *The Victorian Internet: The Remarkable Story of the Telegraph and the Nineteenth Century's Online Pioneers* (New York: Walker; London: Weidenfeld and Nicholson, 1998) is the ideal narrative essay on telegraph technology and its consequences.

BELL AND EDISON

The best way to understand Bell as a person is through *Alexander Graham Bell: The Life and Times of the Man Who Invented the Telephone*, by Edwin S. Grosvenor and Morgan Wesson (New York: Abrams, 1997). It's loaded with photographs from all stages of his life and has long extracts from the love letters as well as good explanations of the basic science. Laszlo Solymar's *Getting the Message: A History of Communications* (Oxford, England: Oxford University Press, 1999) puts the science in the full context of nineteenth- and twentieth-century communication technologies; *Signals: The Science of Telecommunications*, by John Pierce and Noll Michael (New York: Scientific American Library, 1990) neatly brings in information theory as well.

Edison's Electric Light: Biography of an Invention, by Robert Friedel and Paul Israel with Bernard S. Finn (New Brunswick, NJ: Rutgers University Press, 1987) illuminates every question one can possibly imagine about the origin of the electric bulb. For a racily readable account of Edison's life, I'd start with Matthew Josephson's *Edison* (New York and London: McGraw-Hill, 1959), while Neil Baldwin's *Edison: Inventing the Century* (Chicago and London: University of Chicago Press, 1995) takes a more thematic approach. For the shenanigans Edison was involved in when he finally met entrepreneurs as determined as he was, try Jill Jonnes's well-written *Empires of Light: Edison, Tesla, Westinghouse, and the Race to Electrify the World* (New York: Random House, 2003).

For broader trends, Thomas P. Hughes's *American Genesis: A Century of Invention and Technological Enthusiasm 1870–1970* (New York: Viking, 1989) is very good, as is Arnulf Grübler's *Technology and*

Global Change (Cambridge, England: Cambridge University Press, 1998), which has enough charts to make any technological determinist happy; the edited volume by Merritt Roe Smith and Leo Marx's *Does Technology Drive History? The Dilemma of Technological Determinism* (Cambridge, MA: MIT Press, 1996) is a useful counterweight. On electric streetcars, amusement parks, and popular culture generally, see David Nye's *Electrifying America: Social Meanings of a New Technology* (Cambridge, MA: MIT Press, 1990). The collective volume *Science, Technology and Everyday Life 1870–1950*, edited by Colin Chant (London: Routledge, in association with the Open University, 1989) gives a European perspective to go with it. For the delays in taking up the new technology of electrification at the start of the twentieth century, Paul David's writings give a good comparison with the difficulties in getting computer investments to fundamentally change working practices at the end of the same exhausting century. See the discussions in *New Frontiers in the Economics of Innovation: Essays in Honor of Paul David* (Cambridge, MA: Edward Elgar, 2005).

J. J. Thomson's life is sensitively recounted by his son in *J. J. Thomson and the Cavendish Laboratory*, by George Paget Thomson (London: Nelson, 1964), which shows this quiet, seemingly bumbling man ever so thoroughly transforming the modern world; see also Thomson's own autobiography, *Recollections and Reflections* (London: G. Bell & Sons, 1936). What exactly J. J. Thomson discovered in 1897 is an intriguing question, for he couldn't see electrons and was only able to record data on the ratio between electric charge and the mass of "something" in his cathode–ray tubes—a ratio that Walter Kaufmann in Germany had already measured, and with even greater accuracy. Per F. Dahl's *Flash of the Cathode Rays: A History of J. J. Thomson's Electron* (Philadelphia: Institute of Physics Publishing, 1997) is an exciting, narrative-driven account of J. J. Thomson's success; for the general issues of what constitutes discovery, try the essays in *Histories of the Electron: The Birth of Microphysics*, edited by Jed. Z. Buchwald and Andrew Warwick (Cambridge, MA: MIT Press, 2001). Steven Weinberg's *The Discovery of Subatomic Particles* (New York: W. H. Freeman, 1990) is a clearly written synthesis of greater scientific power, and an excellent way to understand Thomson's views, as well as recapping the relation between static and dynamic charges in electromagnetic theory today.

FARADAY

After researching Faraday's life and works for a previous book, I'd thought I had little more to discover, but then I came across James Hamilton's *Faraday: The Life* (London: HarperCollins, 2002) and, boy, did I learn what humility means. Hamilton is superb. He's another art historian, who brings a richness to understanding Faraday's life that most historians of science have missed. We see the tactility in Faraday's hours of quietly winding electrical wire, as well as the significance of touch and smell and the years of deep religious thought, and how it all led to his great discoveries.

Of the earlier accounts, L. Pearce Williams's *Michael Faraday* (London: Chapman and Hall, 1965) is very good; so too is Geoffrey Cantor's *Michael Faraday, Sandemanian and Scientist: A Study of Science and Religion in the Nineteenth Century* (London: Macmillan; New York: St. Martin's Press, 1991). Russell Stannard's interviews with leading scientists, *Science and Wonders: Conversations About Science and Belief* (London and Boston: Faber and Faber, 1996) is a strong start to the general topic of religion and science.

Faraday's key idea was that of a "field." The ideal way to understand it is to spend a summer going through an introductory textbook on calculus, or even just some vector algebra, which would allow the pleasure which is volume 2 of Richard Feynman's *Lectures on Physics* (Reading, MA: Addison–Wesley, 1963; much reprinted) to be accessible. It's hard to look at the world in the same way after that, but as it's also hard to find the free time for the necessary mathematical preparation, an excellent alternative is Einstein's own charming and nontechnical explanation, in parts 2 and 3 of his *The Evolution of Physics, from Early Concepts to Relativity and Quanta*, cowritten with Leopold Infeld (New York: Simon & Schuster, 1966; original edition 1938). The section where Einstein shows the reality of the field by explaining why sparks form when we quickly tug a plug from its socket is especially good. (Chapters 2 and 14 in volume 1 of Feynman's *Lectures* also give a nontechnical flavor of the field approach.)

It's tempting to project back to Faraday's time what we now know of fields, but that would miss the difficulty with which Faraday groped forward to his final achievement. The stages are well explored in P. M. Harman's *Energy, Force and Matter: The Conceptual Development of Nineteenth-Century Physics* (Cambridge, England: Cambridge University Press, 1982), and also in Nancy Neressian's brief yet sharp article "Faraday's Field Concept," in *Faraday Rediscovered*, edited by David Gooding and Frank A. J. L. James (London: Macmillan, 1985; New York: American Institute of Physics, 1989).

Alice Jenkins's essay "Spatial Imagery in Nineteenth-Century Representations of Science: Faraday and Tyndall," in *Making Space for Science: Territorial Themes in the Shaping of Knowledge*, edited by Crosbie Smith and Jon Agar (London: Macmillan; New York: St. Martin's Press, 1998), 181–92, does perhaps the finest job of all at suggesting how visions of three-dimensional landscape entered into Faraday's views, as he moved from imagining instantaneous lines of force to a view of undulating waves traveling at finite speed.

ATLANTIC CABLE, WILLIAM THOMSON AND JAMES CLERK MAXWELL

Charles Bright, the imaginative chief engineer aboard the cable ships, was somewhat less imaginative when it came to naming his own son: he called him Charles Bright. It was the younger Charles who wrote *The Story of the Atlantic Cable* (London: George Newnes Ltd., 1903), and his years of hearing his father's stories make it a great account of the venture; it also has a long transcript from the London *Times* correspondent who was aboard the *Agamemnon* when the HUGE waves of the 1858 storm hit. Bright junior's companion volume, *The Life Story of Sir Charles Tilston Bright* (London, Constable & Co., 1908), is much longer, and contains such detail as the way British sailors would relax while taking samples of the mid-Atlantic seafloor: a crate of beer would go down with the line, at least part of the way, and then be towed up nicely chilled.

Bright tried to be polite about Whitehouse, the electrician who nearly destroyed the venture, but if you're near a good library and can get the official British inquest, *Report of the Joint Committee on the*

Construction of Submarine Telegraph Cables (British Parliamentary Papers 1860, LXII), you're in for a treat, as the spectacle of seeing Victorian barristers letting rip is as impressive in our day as it must have been in theirs. Of recent portrayals, *The Atlantic Cable*, by Bern Dibner (New York: Blaisdell, 1964), and *A Thread Across the Ocean*, by John Steele Gordon (New York: Walker & Co., 2002), both capture the tone well.

My favorite biography of William Thomson (later Lord Kelvin) is that by his friend Silvanus P. Thompson, *The Life of William Thomson: Baron Kelvin of Largs* (London: Macmillan, 1910; two volumes). A great deal more analysis, though perhaps less warmth of understanding, is in *Energy and Empire: A Biographical Study of Lord Kelvin*, by Crosbie Smith and M. Norton Wise (Cambridge: Cambridge University Press, 1989). Crosbie Smith's later *The Science of Energy: A Cultural History of Energy Physics in Victorian Britain* (London: Athlone Press, 1998) does better at getting into the mindset of the time. The way Thomson went beyond Faraday in exploring what was happening inside the cable is the topic of Bruce Hunt's most useful article "Michael Faraday, Cable Telegraphy and the Rise of Field Theory," *History of Technology* 13, (1991), edited by Graham Hollister-Short and Frank A. J. L. James.

It's a shame that the structure of this book grants Maxwell only a brief mention, and I heartily recommend that readers look further into this cautious, brilliant, word-swirling man. Two good biographies are *James Clerk Maxwell: Physicist and Natural Philosopher*, by C. S. F. Everitt (New York: Scribners, 1976), and *The Demon in the Aether: The Story of James Clerk Maxwell* (Edinburgh: Paul Harris Publishing; with Adam Hilger, Bristol, 1983). For Maxwell's key step in using the measured ratio between electrostatic and electromagnetic units to show that light itself is an electromagnetic wave, Simon Schaffer's article "Accurate Measurement *Is* an English Science" in *The Values of Precision*, edited by M. Norton Wise (Princeton, NJ: Princeton University Press, 1995), is a typically lucid introduction.

Both Thomson and Maxwell succeeded by knowing when to construct heuristic physical models, and when to let those models go. For Thomson's bold, simplifying use of Fourier's heat equations in his analysis of the Atlantic cable, see "Mathematics and Physical Reality in William Thomson's Electromagnetic Theory," by Ole Knudsen, in *Wranglers and Physicists*, edited by P. M. Harman (Manchester, England: Manchester University Press, 1985); for Maxwell's intricate construct of spinning gears and idler wheels to help clarify his final 1860s vision (and the edge it gave him over more primly abstract French researchers), start with *The Fire Within the Eye: A Historical Essay on the Nature and Meaning of Light*, by David Park (Princeton, NJ: Princeton University Press, 1997), chapter 9, and then consider Daniel Siegel's *Innovation in Maxwell's Electromagnetic Theory* (Cambridge, England: Cambridge University Press, 1991). After that, Maxwell's own quirkily profound essays become accessible; they're conveniently available in anthologies such as *Physical Thought: An Anthology*, edited by Shmuel Sambursky (London: Hutchinson, 1974).

HERTZ

The diary entries I used in the Hertz chapter are just a portion of what's available in the compelling *Heinrich Hertz: Memoirs, Letters, Diaries*, arranged by Johanna Hertz; second enlarged edition prepared by Mathilde Hertz and Charles Susskind; with a biographical introduction by Max von Laue (San Francisco: San Francisco Press, 1977). Hertz's main scientific papers were collected shortly after his death, and the second edition has an interesting preface by William Thomson; see *Electric Waves: Being Researches on the Propagation of Electric Action with Finite Velocity Through Space* by Heinrich Hertz (London: Macmillan and Co., 1900). Sir Oliver Lodge's *Talks About Wireless* (London: Cassel and Company Ltd., 1925) is a set of vivid reminiscences by someone close to all the main protagonists in radio's early days. The science of what Hertz was doing is covered alongside Maxwell's work in every introductory university text. *The Sciences: An Integrated Approach,* by James Trefil and Robert M. Hazen (New York: John Wiley, 1998), is an easy start, much more fluently written than the norm; Bryan Silver's *The Ascent of Science* (New York and Oxford: Oxford University Press, 1998), takes a more literary approach, dense in historical byways.

RADAR

Watson Watt's own account of his adventures in radar is, as might be expected, very rambling and very compelling; it's *The Pulse of Radar* (New York: Dial, 1959). R. V. Jones's remembrance of the intelligence wars takes a crisp, modestly ironic tone throughout: *The Wizard War: British Scientific Intelligence 1939–1945* (New York: Coward, McCann & Geoghegan, 1978). For biographies of other main characters, C. P. Snow's *Science and Government* (Cambridge, MA: Harvard University Press, 1961) is an excellent start; on Tizard and Lindemann, though, read Solly Zuckerman's *From Apes to Warlords* (London: Hamilton, 1978) as a partial corrective alongside Snow's work; also, see *Tizard*, by Ronald Clark (London: Methuen, 1965). After the official government history *The Strategic Air Offensive Against Germany, 1939–1945*, by Charles Kingsley Webster and Noble Frankland (London: HMSO, 1961; four volumes) came out, Snow revised his 1961 lectures, producing *Appendix to Science and Government* (Cambridge, MA: Harvard University Press, 1962); the back-and-forth makes the arguments, so soon after World War II, come alive once more.

Robert Hanbury Brown's *Boffin: A Personal Story of the Early Days of Radar, Radio Astronomy and Quantum Optics* (Bristol, England: Hilger, 1991) is an unusually well-written memoir, showing the early days of designing the Chain Home system, including the maneuvers needed to protect highly imperfect early equipment against the over-diligent eyes of London inspectors. Jack Nissen's *Winning the Radar War*, co-written with A. W. Cockerill (New York: St. Martin's, 1987), is what one young Cockney experienced when swept into Watson Watt's development teams, where speed was more important than social class—and Britons from East End teenagers to Home County generals found long discussions within these teams liberating in a way they'd never imagined before.

For detailed topics, George Reid Millar's *The Bruneval Raid* (London: Bodley Head, 1974) is good on the paratroop operation, while *The Battle of Hamburg: Allied Bomber Forces Against a German City in 1943* (London: Allen Lane, 1980) is still the best detailed account of the Hamburg raid. As overall introductions to the military campaign, Sven Lindqvist's *A History of Bombing* (London: Granta Books, 2001) is powerful—I took my staccato initial description of Harris in the main text from him—but to balance it, do consider Max Hastings's *Bomber Command* (London: Joseph, 1979, and later editions), which is more conservative, yet equally well researched and nuanced. As an example of the author's psychological insight, he observes that "Although [Harris] sometimes damaged his own cause by his appalling exaggerations, he understood one prime principle of bureaucratic manoeuvre: that by agreeing to a course of action loudly and often enough in public, it is possible in reality to do something entirely different." The BBC radio drama *Bomber* by Len Deighton (BBC Audiobooks, ISBN 0563552662, produced by Jonathan Ruffle) gives a gripping firsthand feel. Robert Buderi's *The Invention That Changed the World: The Story of Radar from War to Peace* (New York: Simon & Schuster, 1996) is the ideal single-volume account of radar through the whole twentieth century, from the forgotten German engineer who patented a shipboard system in 1904 to the latest research in radio astronomy. He also gives space to the fundamental contributions of MIT's Rad Lab, which began to dominate radar development soon after the Chain Home system's great success in 1940. The way one wealthy individual helped jump-start that powerful laboratory is the topic of Jennet Conant's biography of Alfred Loomis, in *Tuxedo Park: A Wall Street Tycoon and the Secret Palace of Science That Changed the Course of World War II* (New York: Simon & Schuster, 2002).

For the underlying science of radar, every introductory text will discuss the free-electron model of metals; Richard P. Feynman's *QED: The Strange Theory of Light and Matter* (Princeton, NJ: Princeton University Press, 1985) uses his wise-guy tone to show how quantum electrodynamics can explain in far more accurate detail what really happens when radar pulses and metals meet.

TURING AND COMPUTERS

The book that restored Turing's reputation was *Alan Turing: The Enigma*, by Andrew Hodges (London: Vintage, 1993; original edition 1983). Even more than the Hamilton biography of Faraday, it's written at a level that far surpasses the usual run of science histories: Hodges is a literary craftsman of the highest rank. He's particularly good at showing how the very concept of transitory, deftly shifting software could have come from Turing's years of having to pass as heterosexual, in a world wary of gays. The key years at Princeton and the importance of Turing's sexuality are a running theme in Neal Stephenson's terrific page-turner of a novel, *Cryptonomicon* (New York: Avon, 1999). An ideal account of the Bletchley Park days is Simon Singh's *The Code Book* (London: Fourth Estate, 1999), in which the details of cryptography become as compelling to the reader as they were to Turing.

Mathematically inclined readers will note that since ciphers can be applied and reversed in sequence, they follow the rules of group theory from abstract algebra—which is why mathematicians receive high salaries at GCHQ and the NSA. Israel Herstein's perennial *Topics in Algebra* (Lexington, MA: Xerox College Publications, 1975) suggests what sophisticated codebreaking techniques this permits, and gives a feeling for the beauty that abstraction can give. Constance Reid's tenderly written biography *Hilbert* (New York: Copernicus, 1996; original edition 1969), based on interviews with many of Hilbert's colleagues, is the ideal nontechnical introduction to the seemingly fearsome *Entscheidungsproblem*. *Insights of Genius: Imagery and Creativity in Science and Art*, by Arthur I. Miller

(Cambridge, MA: MIT Press, 2001), gracefully explicates the links that allow a Turing to see things no one had before.

For the early history of computing, *Glory and Failure: The Difference Engines of Johann Müller, Charles Babbage, and George and Edvard Scheutz*, by Michael Lindgren (Cambridge, MA: MIT Press, 1990) has a fun look at the two Swedish engineers who actually built Charles Babbage's Difference Engine No. 1—but then found, in the 1840s, that no one was ready to buy it. *Turing and the Universal Machine: The Making of the Modern Computer*, by Jon Agar (Cambridge, England: Icon Books [published in the U.S. by Totem Books], 2001) points out the way that the growth of social security programs and large bureaucracies in general made computers almost inevitable. *The Dream Machine: J. C. R. Licklier and the Revolution That Made Computing Personal*, by M. Mitchell Waldrop (London and New York: Viking, 2001), shows how the need to personalize the serried ranks of 1950s missile defense consoles led to the personal computer. *Machines Who Think: A Personal Inquiry into the History and Prospects of Artificial Intelligence*, by Pamela McCorduck (Natick, MA: A.K. Peters, 2003; updated reissue of 1979 W. H. Freeman edition), brings alive the first generations of researchers who carried on, often with more enthusiasm than success, Turing's dream of artificial intelligence.

Anyone who's struggled to make large computer systems work—and anyone who's wondered why it's so hard—will enjoy my favorite of all books from a working system designer: *The Mythical Man-Month: Essays on Software Engineering* (Reading, MA: Addison–Wesley, anniversary edition 1995); it's brought up to date by Eric Raymond's bardlike collection *The Cathedral and the Bazaar: Musings on Linux and Open Source by an Accidental Revolutionary* (Sebastopol, CA: O'Reilly, 1999). Georgina Ferry's *A Computer Called Leo:*

Lyons Teashops and the World's First Office Computer (London: Fourth Estate, 2003) is a plucky look at the not–quite–successful early British efforts to create a computer industry. Danny Hillis's *The Pattern on the Stone* (London: Weidenfeld and Nicholson, 1998) is the perfect brief introduction to the science of computing; Neil Gershenfeld's *The Physics of Information Technology* (New York: Cambridge University Press, 2000) is a more advanced look at how it's done.

TRANSISTORS AND QUANTUM MECHANICS

Michael Riordan and Lillian Hoddeson interviewed a large number of the participants from the race to create a working transistor, which gives their *Crystal Fire: The Invention of the Transistor and the Birth of the Information Age* (New York: W. W. Norton, 1997) a great immediacy and authority. Hoddeson took a similar approach in her biography of the modest theorist who led the Bell Labs team, *True Genius: The Life and Science of John Bardeen*, coauthored with Vicki Daitch (Washington, DC: Joseph Henry Press, 2001). Both books show how a technology that seems obvious in retrospect can be cloudy and tentative at the moment it's being born.

The underlying science is fundamental to every quantum mechanics text; two good examples are *Quantum*, by Jim Al-Khalili (London: Weidenfeld & Nicolson, 2003), and the somewhat more historically based *The Quantum Universe*, by Tony Hey and Patrick Walters (Cambridge, England: Cambridge University Press, 1987), which has such tidbits as an account of the unfortunate British radar engineer G. W. A. Dummer, who in 1952 came up with detailed plans for an integrated circuit—yet ended up being ignored, owing to the same administrative stodginess that had crushed

Turing. His plans were independently developed in America to immense success nearly a decade later.

John Polkinghorne's *Quantum Theory: A Very Short Introduction* (Oxford and New York: Oxford University Press, 2002) is a stylish summary by a longtime Cambridge don, including the best brief recap I know of band structure within crystalline solids. Lawrence Krauss's *Fear of Physics: A Guide for the Perplexed* (New York: Basic Books, 1994) is a playful yet sharp survey of the range of modern physics.

Rudolf Peierls gives a personal account of the key concept of "holes" in his whimsical, wise *Bird of Passage: Recollections of a Physicist* (Princeton, NJ: Princeton University Press, 1985). More-systematic analyses of each stage in this new science's development are in *Out of the Crystal Maze: Chapters from the History of Solid State Physics*, edited by L. Hoddeson, E. Braun, J. Teichmann, and S. Weart (New York: Oxford University Press, 1992). Readers with even a little German will enjoy exploring the wild psychological roots of Pauli's conception of the Exclusion Principle, in *Wolfgang Pauli und C. G. Jung. Ein Briefwechsel 1932–1958*, ed. C. A. Meier (Berlin and Heidelberg: Springer, 1992).

Two important technologies that didn't have a place in this book are well introduced in *Tube: The Invention of Television*, by David E. Fisher and Marshal Jon Fisher (Washington, DC: Counterpoint, 1996), and *How the Laser Happened*, by Charles Townes (New York and London: Oxford University Press, 1999). Aside from modestly recalling the quiet morning, waiting on a park bench in Washington, DC, when he first came up with the idea for a laser, Townes also describes his invention's applications in industry, as well as the giant lasers that have formed naturally in outer space. The mix of billions of years of time, low density of matter, and massive available energy sources produced the ideal conditions for these beams to light up.

RESEARCH LABS AND THEIR CONSEQUENCES

When you work at an organization that developed yet failed to profit from the mouse, the concept of digital folders, scrolling, pointing and clicking, and indeed almost all key properties of the personal computer, you get rather attuned to issues of a research laboratory's failures and successes. This is perhaps why John Seely Brown, long-term director of Xerox PARC (where those failures occurred, though before his tenure) has written, with Paul Duguid, the very best account of how innovation in companies and in society actually operates: *The Social Life of Information* (Boston: Harvard Business School Press, 2000).

To go with that, *Up the Infinite Corridor: MIT and the Technological Imagination*, by Fred Hapgood (Reading, MA: Addison-Wesley, 1993), is a calm anthropology of that strange, world-transforming creature known as the MIT engineer; Richard Rhodes's edited anthology *Visions of Technology: A Century of Vital Debate About Machines, Systems and the Human World* (New York: Simon & Schuster, 1999) samples the discussions about what they've achieved. Peter Hall's *Cities in Civilization: Culture, Innovation, and Urban Order* (London: Weidenfeld & Nicolson, 1998) belies its massive size: it's a wonderfully readable account of how societies change, with a long chapter on creativity in Elvis Presley's Memphis, Tennessee—right at the moment when transistor technology was about to arrive. *The Human Web: A Bird's-Eye View of World History*, by John R. McNeill and

William H. McNeill (New York: W. W. Norton, 2003), puts information transformations at center stage in the history of civilization; in my estimation their book will be seen as a seminal guide to what our twenty-first century has to offer.

THE MIND AND BEYOND

There's excellent background on every area of physiology in Colin Blakemore and Sheila Jennett's beautifully illustrated *The Oxford Companion to The Body*, edited by Colin Blakemore and Sheila Jennett (Oxford and New York: Oxford University Press, 2001). Hodgkins's own detailed account of the state of play two decades after his and Huxley's main breakthrough is *The Conduction of the Nervous Impulse* (Liverpool: Liverpool University Press, 1963); even more telling for his own experiences is the quiet, long autobiography he wrote in 1992: *Chance and Design* (Cambridge, England: Cambridge University Press). For the brain overall, Susan Greenfield provides a schoolmistressly brisk account in *The Private Life of the Brain* (New York: John Wiley, 2000); John McCrone's *Going Inside: A Tour Round a Single Moment of Consciousness* (London: Faber and Faber, 1999) is a more easygoing, detailed exploration. David Hubel's *Eye, Brain and Vision* (San Francisco: Scientific American Library, 1988) is a fine vision of what seeing entails, presented by the scientist who did much to reveal it. He also writes well, forgoing the usual phrases about ions transferring across membranes, for the vividness of "little machine-like proteins . . . which seize ions and eject them from the cell."

A beautiful intellectual history is *From Beast-Machine to Man-Machine: Animal Soul in French Letters from Descartes to La Mettrie*, by

Leonora Cohen Rosenfield (New York: Octagon Books, 1968), which shows, in appropriately teleological fashion, how the intellectual preconceptions the nineteenth–century physiologists needed came about. P. Cranefield's article "The organic physics of 1847 and the biophysics of today" in *The Journal of the History of Medicine and Allied Sciences* 12 (October 1957) goes into the seminal, semi-secret meeting in Berlin in 1847 when four young scientists decided it was time to break from all prior religious authority and measure the finite speed of electric currents within nerves; Leo Koenigsberger's biography *Hermann von Helmholtz: A Life*, translated by F. A. Welby (Oxford, England: Oxford University Press, 1906) shows the tension the most important thinker of that group felt as their project was about to begin. Thomas Kuhn's essay "Energy Conservation As an Example of Simultaneous Discovery" in *The Essential Tension: Selected Studies in Scientific Tradition and Change* (Chicago and London: University of Chicago Press, 1977) puts their work in context, showing that from the moment of Faraday's 1831 experiment (which finally disproved Volta's vague contact theory of electrical generation), the stage was set for at least half a dozen investigators in the subsequent two decades to "independently" discover the conservation of energy. *A Short History of Anesthesia*, by G. B. Rushman et al. (London: Butterworth–Heinemann, 1996), traces scientific physiology infiltrating into the most hidebound of surgical practices.

On the developments in brain receptors that led to our targeted mood–altering drugs, Solomon Snyder's *Brainstorming: The Science and Politics of Opiate Research* (Cambridge, MA: Harvard University Press, 1989) shows the author as a wise, avuncular laboratory chief; the memoir by one of his most disgruntled ex–colleagues,

Molecules of Emotion, by Candace Pert (New York: Simon & Schuster, 1997), presents a somewhat different view. In either case, Snyder's survey *Drugs and the Brain* (San Francisco: Scientific American Library, 1986) is a good sourcebook for the basic operation of synapses and receptors. For what the Prozac revolution means, there's no text better than Peter Kramer's kind, insightful *Listening to Prozac* (London: Fourth Estate, 1994).

Finally, to explore more from the part introductions and epilogue, the late Heinz Pagels's *The Cosmic Code* (New York: Simon & Schuster, 1982) is a poetic masterpiece on the microworld and macroworld, while Steven Weinberg's *The First Three Minutes: A Modern View of the Origin of the Universe* (London: Deutsch, 1977) shows how even simple high school equations, properly understood, are capable of describing the universe's first moments in tremendous detail. Martin Rees's *Our Cosmic Habitat* (Princeton, NJ: Princeton University Press, 2001), and also *The Five Ages of the Universe*, by Fred Adams and Greg Laughlin (New York: Simon & Schuster, 1999) demonstrate the grandeur and perspective that the finest of astrophysicists can provide.

ACKNOWLEDGMENTS

Once again this book grew out of lectures from my years of teaching at Oxford, for which many thanks are owed to Ralf Dahrendorf, Avi Shlaim, and Roger Owen, all then of St. Anthony's College. The idea of turning it into a text that dealt as much with practical technology as with the underlying science came in part from time I spent with research or scenario teams at a number of great enterprises, most especially Microsoft, BMW, Shell, and Pfizer; in part that angle also came from my extended family's laconic attitude toward the world: a feeling that it is an odd and marvelous place, but that with calm understanding—and a respect for practical skills—we can navigate our way within it.

My uncle Murray Alpert, who died just as I was finishing this text, was exceptionally kind in passing that attitude on to me. He was one of the longest serving members of the U.S. Air Force, joining the Army Air Corps shortly before World War II, and finishing in the era of F-117 Stealth fighters. He had a calm attitude to technological jobs, making little distinction between putting in the plumbing for a house's shower on a quiet weekend and supervising the electrical system for one of these massive military jets. I loved watching him work.

The book itself begins with my late father, first as a little boy, and then as a young man moving to Chicago in the 1920s; it ends with my mother, who was the Ohio farm girl who met him shortly after World War II, and with whom she began the family that was around her, nearly sixty years later, when she shared his first love letters. Any insight this book might have comes from what I've learned from them. My mother's two brothers, Len and Gene Passell, and her sisters, Sarah Alpert and Doris Easton, also shared a wonderfully no-nonsense attitude: I remember with great affection, when they were all in their seventies or eighties, watching the television news together at the farmhouse in Ohio and realizing: there is more than four hundred years of accumulated memory here, and I'm getting to share that through their apt, wise words.

A great number of friends helped at various stages of the project, including Rebecca Abrams, Shanda Bahles, Sunny Bates, Julia Bindman, Jasmine Birtles, Doug Borden, Richard Cohen, Esther Eidinow, Janet Evans, Anne and Chris Finn, Betty Sue Flowers, Brandy Freisinger, Matt Golde, Rhonda Goldstein, Joe Hajnal, Tim Harford, Matthew Hoffman, Natasha Illum Berg, Joanna Kalmer, Tara Lemay, Adam Levy, Suzanne Levy, Sue Liburd, Karen Liebrich, Peter Main, Terry Manning, Arthur Miller, Fran Monks, Dan Newman, Mia Nybrant, Teresa Poole, Ramana Rao, Marica Rosengard, Harriet Rubin, Jonathan Ruffle, Tira Shubart, Julia Stuart, Jennifer Sullivan, Ilan Troen and above all Gabrielle Walker. I'd like to thank them all individually, but for once I'm lost for words: all I can note is that their kindness, and insight, transformed the book for the better.

This project was begun at Crown with Emily Loose and ably completed by Rachel Klayman, all under the helpful supervision of Kristin Kiser and Steve Ross; their forbearance in dealing with an author who always seemed to want "just one more change" is most appreciated. Crown is a big firm, but somehow makes authors feel part of a community. I remember one time, chatting with Rachel and Kristin and Steve, and exploring just for fun how Erik Larson put together his books: that led to my turning to *Isaac's Storm*, and from it getting the idea for the italicized framing pages that set the electricity story in a wider astronomical perspective. Without their conversation, I never would have tried that.

The actual writing was begun with the author and his papers spread out on the front seat of a parked Land Cruiser 6,000 feet up in Tanzania, windows rolled up in a vain attempt to block the howling gale up from the Great Rift Valley, while an enthusiastic Scandinavian friend was calling out—it was hard to hear her over the wind's roar—that it wasn't usually this cold! That any moment now the wind would fade! She was right, albeit off by a few hours, and sitting on her balcony a little later, watching the sunset and regaining my circulation, gossiping about this and that, and then half-closing my eyes to listen to her read aloud, I realized what the tone of the book could be. The world is old, but electricity is older. It shaped the Masai Hills I could vaguely see; it's shaped the lives of everyone who's walked across those hills as well.

The book is now being completed, five miles over the Atlantic, jetting in tranquil comfort beside my eight-year-old daughter, home soon to her big brother. How they helped they will only realize later. When there was a lot of writing to do they chimed in with a cascade of eager questions; when I lost focus in any of the chapters, I'd try out fresh stories as we walked, skipped, scooted and otherwise locomoted our way to school in the morning.

I did some of the writing at night but mostly took to getting up very early and doing it then. We have a big kitchen, with big windows, and I loved waking in the dark, with London quiet outside; then puttering to the kitchen, and making coffee or tea before spreading out my papers on the wooden table where we'd later have our breakfast. Usually I'd have stopped to look in on the kids in their rooms—Sam with his giant Homer Simpson cut-out; Sophie with the fairy castles on the wall—but even when I hadn't, I felt a great tranquillity just from knowing they were there.

Gradually the birds would get louder and louder from the gardens outside, and early hints of light would appear; by six-thirty or seven the first sleepy pyjama-clad child would wander in, ready for some chatting or at least emergency supplies of fruit juice or warm chocolate. They'd lean against the wall under the biggest window, either reading their own books or drawing a little, or sometimes not doing much of anything, just happy to keep their dad company. I tried not to smile as I scribbled away, feeling their thoughts, their tenderness, enter my soul.

Feeling their contentment enter my words.

INDEX

Watt, James, 115, 235–36
Watts, 235–36
Waves, 190
  bending of, 110
  electric, 6–7, 147, 149
  electromagnetism, 66–67, 89–91,
    95–108, 197–98
  of extrasensory phenomena, 7,
    263
  invisible, 7, 54–56, 65–70, 91–92,
    110
  light, 174, 248–49
  measurement of, 92, 109
  oscilloscopes in detection of,
    121, 127
  radar, 122–29
  radio, 110, 111–12, 114–21
  reflection of, 102–3
  religious faith in, 7
  in satellite navigation, 187–89
  undersea cable, 73–86
Web searches, 189–90
Western Union, 39, 42
Wheatstone, Charles, 21
Whitehouse, Edward O., 78–80,
  81, 82–83, 84, 85, 86
Wilkins, Arnold, 116–17, 118–20,
  236, 258
Wimperis, Henry, 121
Winkler, Clemens Alexander, 258
Wires:
  birds sitting on, 245
  electric current in, 6, 16–20, 25,
    36, 41, 53, 66, 174–75
  for electric lights, 43–44

electromagnetism in, 66–67
electrons in, 18, 53, 174–75, 190
glass insulators for, 174–75
Henry and, 15–20
and invisible force field, 76
resistance in, 36, 42
Sturgeon and, 16, 17
telegraph, 19–20, 29–30, 77, 200
telephone, 29–32, 34–37, 41
undersea cable, 72, 73–86, 227
Volta and, 16
wall sockets of, 46, 246
World War I, 121, 123, 250
World War II, 121–29
  bombing raids in, 140–48, 251
  codebreaking in, 162–66, 168
  Luftwaffe intelligence reports,
    126
  Munich Crisis, 124
  Operation Sea Lion, 128
  radar in, 122–29, 130–43,
    145–48, 149, 205, 229–30, 236
  unity in, 169
  Wrens in, 163–64, 165
Wrens (Women's Royal Naval
  Service), 163–64, 165
Würzburg radar, 133–41, 146,
  229–30

Yeats, William Butler, 162
Young, Leo, 114

Zeppelins, 125, 229

ABOUT THE AUTHOR

DAVID BODANIS taught a survey of intellectual history at the University of Oxford for many years. He is the author of several books, including the bestselling $E=mc^2$, which was translated into twenty languages, and *The Secret House*. A native of Chicago, he lives in London, England.

3 1142 00727 5001